SPONSORS OF THE EPSC 1995–1996

- AEA Technology
- Air Products plc
- Akcros Chemicals
- Akzo Nobel Engineering bv
- Arco Chemical Europe Inc
- BASF AG
- Bayer AG
- Borealis
- Clariant International Ltd
- The British Petroleum Company plc
- Ciba-Geigy Ltd
- DNV Industry UK Ltd
- The Dow Chemical Company
- DSM
- Du Pont de Nemours International SA
- EC Joint Research Centre — Institute for Systems, Informatics and Safety
- Elf Atochem
- Exxon Chemical Europe Inc
- Hoechst AG
- ICI
- Merck, Sharp & Dohme
- Monsanto
- Norsk Hydro as
- Procter and Gamble — European Technical Center
- Rhône-Poulenc
- Rohm & Haas (UK) Ltd
- Sandoz Technology Ltd
- Schering-Plough (Avondale) Company
- Shell International Chemicals BV
- Solvay SA
- TNO
- TÜV Südwest
- VTT Manufacturing Technology
- Zeneca

EPSC BENEFITS

- Improved cross-European co-ordination on safety standards.

- Identification of areas where manuals and guidelines could be produced.

- Improved co-ordination of safety R&D and handling of complex technical research programmes.

- Stimulation of R&D in areas where there are gaps in knowledge.

- Transfer of knowledge from elsewhere to Europe and between European countries.

- Technical input to legislators and standard makers to ensure more realistic legislation.

- Sharing and dissemination of information on safety technology and accident prevention.

- Access to information from a single source.

SAFETY PERFORMANCE MEASUREMENT

Edited by
Jacques van Steen, TNO

for the European Process Safety Centre

Published by the Institution of Chemical Engineers, Davis Building,
165-189 Railway Terrace, Rugby, Warwickshire, CV21 3HQ, UK.
A Registered Charity

Copyright © 1996 European Process Safety Centre

ISBN 0 85295 382 8
Reprinted 1997

Typesetting and illustrations by Sheffield Academic Press Ltd,
Mansion House, 19 Kingfield Road, Sheffield, S11 9AS, UK

Printed in the United Kingdom by Bookcraft Ltd, Bath, Somerset

Preface

This publication was compiled from contributions of representatives from the European Process Safety Centre (EPSC) subcommittee on Safety Management Systems. It meets the EPSC's first objective — to provide information on currently accepted process safety practices. As such, it is a logical sequel to the EPSC publication 'Safety management systems' which appeared in 1994.

The EPSC is an international industry-funded organization which exists to provide an independent technical focus for process safety in Europe. It is a cooperative project administered by the Institution of Chemical Engineers on behalf of the European Federation of Chemical Engineering (EFCE).

The EPSC subcommittee on Safety Management Systems consists of the following members:

B Fröhlich (Chairman)	Exxon Chemical Company
I Dalling	AEA Technology
P Ruckh	BASF AG
G C W Walker	BASF plc
W Viefers	Bayer AG
J-H Christiansen	Borealis
R G Read	The British Petroleum Company plc
F Stoessel	Ciba-Geigy AG
G Suter	CLARIANT International
W N Top	DNV
R Gowland	The Dow Chemical Company
W J Kolk	Du Pont de Nemours (Nederland) BV
E Guntrum	Hoechst AG
J L Hawksley	ICI
B Hotson	Monsanto plc
M H Brascamp	Netherlands Organization for Applied Scientific Research TNO
W Bjerke	Norsk Hydro as
P G D van der Want	Shell International Chemicals BV
C Bartholomé	Solvay SA
G Burgbacher	TÜV Energie und Umwelt GmbH
N Byrom (by invitation)	UK Health & Safety Executive
B M Hancock (Secretary)	European Process Safety Centre

This book was published in conjunction with a one-day international EPSC conference on current practices and future developments in safety performance measurement (Paris, 18 October 1996). Contributions for this publication were provided by representatives of fourteen EPSC member companies.

Contents

1 Introduction

1.1 Background

Recently the European Process Safety Centre (EPSC) prepared the book 'Safety management systems', published by the IChemE in 1994. That publication was compiled from contributions of representatives from the EPSC subcommittee on safety management systems, given in discussions with DG XI of the Commission of the European Communities. Its purpose was to share with others the experiences gained at the time by EPSC member organizations on safety management and safety management systems. Thus, it went beyond the level of generic safety management system concepts and was mainly concerned with presenting typical examples from industry.

The subcommittee has now turned its attention from safety management systems to safety performance measurement. This started as an internal benchmarking process which provoked sufficient interest for it to be taken into the public domain. This book is a logical sequel and a companion volume to 'Safety management systems'. It broadens the issue of implementing safety management systems by adding a perspective on measuring their performance. The eventual goal of this book is to demonstrate the benefits and increase the use of performance measurement in the process industries. More specifically, its purpose is to:

● clarify the meaning and potential of safety performance measurement;
● show the breadth of techniques and approaches in measuring safety performance.

This book begins with a generic framework for performance measurement, but its emphasis is on current, and developing, practice in industry. Thus, in line with the character of 'Safety management systems', it also focuses on real examples.

This book is not a textbook on safety performance measurement or a manual for measuring performance. It is primarily a broad collection of specific examples from specific companies, although these examples are positioned within a generic framework in order to put them in the right perspective. It must be emphasized that designing and implementing safety performance measurement in a particular company cannot just be copied from other companies' systems but has to be tailored to the specific needs and characteristics of the company in question.

1.2 Structure of the book

This book is organized as follows. Chapter 2 provides a general introduction to safety performance measurement. It discusses some key concepts of safety performance and presents a framework for performance measurement. Chapters 3 to 6 form the main part of the book. These contain specific examples on measuring safety performance, which are positioned within the framework.

2 Safety performance measurement

This chapter contains a general introduction to safety performance measurement. It begins with discussing some key concepts of safety performance, and goes on to present a framework for performance measurement. Since many organizations are integrating their systems for safety, health and environment protection, the same concepts and framework apply to a large extent to combined safety, health and environment performance measurement.

The practical examples which form the main part of the book are positioned within the framework.

This chapter draws heavily on an article by J L Hawksley in *The Chemical Engineer*, 25 April 1996, published by the Institution of Chemical Engineers.

2.1 Concepts and framework

Safety can be defined as the absence of danger from which harm or loss could result. Then the only direct measure of achieved performance is in terms of the harm or loss that does occur, and reducing losses provides direct evidence of performance improvement. That is, success can be measured through absence of failures, and so injuries, illnesses, losses etc have to be measured — they are the 'bottom line' of safety performance.

Measuring outputs is characterized by two important limitations:

1.   When safety is good and injury and loss rates are low, then those measurements are not sufficient to provide adequate feedback for managing safety.
2.   For operations where there may be potential for severe accidents, the likelihood of such events must be extremely low. This means that the absence of very unlikely events is not, of itself, a sufficient indicator of good safety management.

Therefore, other — proactive — measurements of safety performance are necessary. There need to be indicators which give assurance that the absence or reduction of harm or loss is due to a systematic management approach which is aimed at preventing the occurrence of incidents. This approach contrasts with the reactive management approach, which initiates actions and programmes after undesired events.

2.1.1 Safety management inputs

Safety is assured by providing:

• plant and equipment which is 'fit for the purpose' of reducing the risks from identified hazards as far as is reasonably practicable;

● systems and procedures to operate and maintain that equipment in a satisfactory manner and to manage all associated activities;

● people who are competent, through knowledge, skills, and attitudes, to operate the plant and equipment and to implement the systems and procedures.

These are the positive inputs of safety management which are put in place to prevent the negative outputs (the failures). There are two maxims that are worth noting: 'Accentuate the positive to eliminate the negative' and 'You cannot manage what you do not measure'. This means that performance indicators are also required for the positive inputs. So safety performance measurement has to cover four areas: three that are essentially positive and one that is essentially negative (see Figure 2.1). Continual improvement in safety management is about proactively expanding the positive inputs to reduce the negative outputs — that is, to reduce the total of incidents which create harm and loss to people, environment and assets (see Figure 2.2). This will enable the safety management system effort to improve continuously in effectiveness and efficiency, thereby controlling and reducing the risks of the operations.

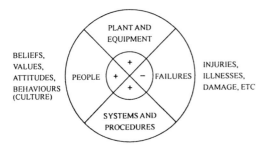

Figure 2.1　*Safety performance measurement — the areas to be covered*

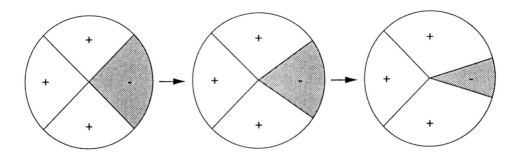

Figure 2.2　*Safety performance — continual improvement*

2.1.2 *A framework for performance measurement*

The implementation of the inputs to safety management can be monitored by a variety of inspections, assessments and audits. These provide the means for positive measures of performance. The monitoring activities fall into three general categories:

● Regular, and often frequent, inspections and audits carried out mainly by local management and staff. The frequency may be daily, weekly, monthly or less, depending on what is being monitored.

● Periodic, 'in-depth' inspections, assessments or audits of some specific aspect. These are carried out mainly by independent specialists together with local staff, typically at intervals of several years.

● Overview assessments and audits. These are conducted by independent experienced assessors, typically at intervals of one to three or more years.

The independent specialists or assessors may be 'in-house' personnel independent of the facility being monitored, or they may — if appropriate — be external, 'third party' personnel.

There should be some monitoring activity covering each of the above categories for each area of management input (plant, systems, people): each sector of the matrix in Table 2.1 should be covered. In practice, the distinction between sectors is not necessarily as definitive — often a particular inspection/assessment/audit activity will simultaneously cover all or part of several sectors of the matrix.

It is crucial that these measurements or indicators are fed back into the management system loop in an understandable form, in order to enable the safety or risk manager to improve the safety management process of the operation.

Table 2.1 *Safety performance measurement — monitoring the inputs*

	Regular inspections and audits by local staff	Periodic in-depth inspections, assessments, audits of specific aspects by specialists	Overview assessments and audits by independent experienced assessors
Plant and equipment	1	5 6 7	10
Systems and procedures	2 3	5 6 7 8	10
People	4	9	10

The numbers in the matrix sectors of Table 2.1 indicate the main areas covered by the following general examples of inspection, assessment or audit activities:

1.   Inspections of equipment and facilities; examination and testing of protective systems and devices; housekeeping inspections.

2.   Compliance audits to verify the implementation of specific systems and procedures, in order to answer the question 'do we do what we say we do?'.

3.   Measures of management control inputs — for example, percentage of improvement actions completed on time, numbers of 'learning events' communicated and discussed.

4.   Behavioural observation and feedback — for example, safe/unsafe behaviours audits

(both workforce behaviours and management behaviours).

5.   Specialist audits — for example, of engineering standards — against stated internal and/or external standards and recognized good practice.

6.   Mechanical integrity examinations and inspections — for example, periodic pressure vessel checks.

7.   Hazard and risk assessments — for example, to indicate the results of hazard and risk reduction improvement.

8.   Systems integrity audits to confirm system robustness and adequacy, in order to answer the question 'is what we say we do good enough?' More than just a 'quality systems' audit, these audits must also test that the system covers all the necessary safety requirements.

9.   Attitude surveys.

10.   Overall 'management' audits to confirm that all requirements are being met. Mostly such overview audits will involve sample checks covering other sectors of the matrix. For example, an overview audit would typically include compliance checks of a number of procedures and would check in some detail the adequacy of a sample of procedures.

It must be emphasized that these are indicative general examples only. Specific monitoring activities as implemented by different organizations may be described in different ways and have differing scopes: they can and will vary from one organization to another. But a key factor is that monitoring has to involve a number of complementary activities. No single monitoring activity can cover all requirements.

    Another key factor is that safety performance cannot be expressed in terms of a single parameter or index. Rather there is a range of mostly qualitative, but sometimes quantitative, indicators from each monitoring activity. The separate measurements can be expressed in a variety of ways. These include:

● inferences based on, for example, the number and nature of defects or nonconformances found, the nature and type of recommendations made;

● qualitative assessments of performance on broad scales from, say, 'poor' (immediate improvement action needed) through 'satisfactory' (capable of improvement) to 'good' (best practice, no action necessary);

● quantitative ratings — for example, percentage compliance with the various specific elements of the management system;

● quantitative and/or qualitative ratings about the quality of the safety management activities and about system implementation commensurate with the risks of the operations.

    'Benchmarking' against another organization (ideally the 'best in class') is also a valuable way of measuring performance across any of the sectors in Table 2.1.

2.2   Introduction to the practical examples

This book is primarily concerned with presenting practical examples of safety performance measurement which have been provided by representatives of European Process Safety Centre member companies. These examples are organized according to the framework already discussed. There are separate chapters for each of the three areas of safety management input (plant and equipment, systems and procedures, and people), followed by a chapter about output measurement.

All authors were asked to focus on a particular category of monitoring activities within a specific area of safety management input, and to write self-contained sections which discuss their own company's current practice in that area. The result illustrates some of the statements on performance measurement made in section 2.1:

● The boundaries between the matrix sectors of Table 2.1 are not necessarily sharp: in practice, monitoring a particular sector often goes together with monitoring adjacent sectors.
● How particular monitoring activities are implemented can and will vary between different companies.

Thus, the following chapters provide no more and no less than specific examples from specific companies. Each is only a part of the overall system of the company. Because they derive from different overall systems, however, they do not together constitute an overall system of monitoring and measurement. Moreover, as indicated before, designing and implementing a safety management system and the associated measurement of its performance in a particular company cannot just be copied from other companies' systems but has to be tailored to the specific needs and characteristics of the company in question.

Editor's note
This book is primarily an edited collection of contributions within a generic framework. Whereas the individual contributions as submitted originally differed strongly in character, the editor's goal has been to preserve those differences as much as possible and at the same time to present the resulting compilation as a comprehensive whole. Thus, editorial emphasis has been placed on achieving a coherent book as well as consistency in style, layout and presentation whilst maintaining the character of the individual contributions.

3 Measuring plant and equipment: practical examples

The first area of safety management inputs is concerned with plant and equipment, which should be 'fit for the purpose' of reducing the risks from identified hazards as far as is reasonably practicable (see Table 3.1). Monitoring activities in this area include technical inspections, specialist audits against technical standards and good practice, hazard and risk assessments, and overall management audits.

Table 3.1 *Safety performance measurement — monitoring the inputs*

	Regular inspections and audits by local staff	Periodic in-depth inspections, assessments, audits of specific aspects by specialists	Overview assessments and audits by independent experienced assessors
PLANT AND EQUIPMENT			
Systems and procedures			
People			

This chapter presents five examples of monitoring activities in the plant and equipment area. To begin with, section 3.1 describes integral plant inspection at Bayer. The inspections in question take place in a three-level system in which the operator, independent experts and the authorities all have their role. The emphasis of the section is on the first two levels. An overview is given of the various inspections which are carried out by the operator. These include inspections by the plant operator and relevant departments as well as by so-called operator officers for special tasks. Section 3.1 then gives an impression of certain inspections which have to be carried out by independent experts.

In section 3.2, the integrated modular audit system at Hoechst is discussed. The idea behind this system is to incorporate as many different audit aspects as possible under one roof. The system contains 18 modules. Each module consists of a standard procedure which describes the audit process in detail. The audits are mainly concerned with identifying deviations which are associated with processes, organization and equipment/documentation. The section includes examples of the different types of deviations, and also a discussion of practical aspects and experiences of applying the system.

The contribution from TÜV Südwest (section 3.3) is concerned with technical audits and inspections from a more general perspective. Auditing safety management systems is discussed particularly in the context of the new Seveso II Directive. Audit techniques as well as audit tools and methods are briefly described. The section concludes by presenting the results of an actual safety management system audit in terms of Seveso II requirements.

Section 3.4 by DNV addresses condition measurement. This involves at least two measurements: the measurement of physical conditions and the measurement of risk. The evaluation of physical conditions, which is based on carrying out observations, is briefly described. The emphasis of the section is on DNV's risk-based inspection (RBI) approach. RBI risk measurement, the analysis of its results, and the associated follow-up are discussed. The section includes an example from an RBI study in a petro-chemical unit.

Finally, section 3.5 concerns the development of a safety performance measurement system at Borealis. It covers various aspects of measuring safety performance: injury performance measurements, auditing HSE performance and surveying operational safety. The emphasis is on the latter two. The auditing process as well as the operational safety survey are described. The section includes a process hazard checklist and an example of a checklist for process standards which are used in the operational safety survey.

The various types of monitoring activities in the plant and equipment area, as described in section 2.1, are covered in this chapter. Although the individual sections were intended to focus on a particular category of monitoring activities within that area, they may take a wider perspective to greater or lesser extents. This demonstrates that the boundaries between the matrix sectors of Table 3.1 are not necessarily sharp: in practice, monitoring a particular sector often goes together with monitoring adjacent sectors. Moreover, there may also be overlaps with monitoring health and environment issues within integrated systems for safety, health and environment protection.

3.1   Integral plant inspection
W Viefers, Bayer

An important element for ensuring comprehensive plant safety in Germany is integral plant inspection, as described in *Leitfaden Anlagensicherheit, (Störfall-Kommission beim Bundesminister für Umwelt, Naturschutz und Reaktorsicherheit)*, November 1995. Integral plant inspection provides for examinations throughout a plant's entire life span, i.e. planning, manufacturing and erection, commissioning, operation, decommissioning, closure, demolition and disposal. Acts and ordinances from all relevant areas of law serve as the basis, as well as the requirements resulting from official licensing procedures (see Figure 3.1, pages 10 and 11). The tests and inspections which are required monitor the proper functioning of individual apparatus and systems as well as the entire plant. Within an overall concept of performance measurement of the safety management system in use, they form base-line activities assuring the safe technical state of the plant. The implementation of the regulatory requirements at Bayer AG is described below.

Integral inspection takes place in a three-level system, in which operators, experts and authorities are allocated different tasks:

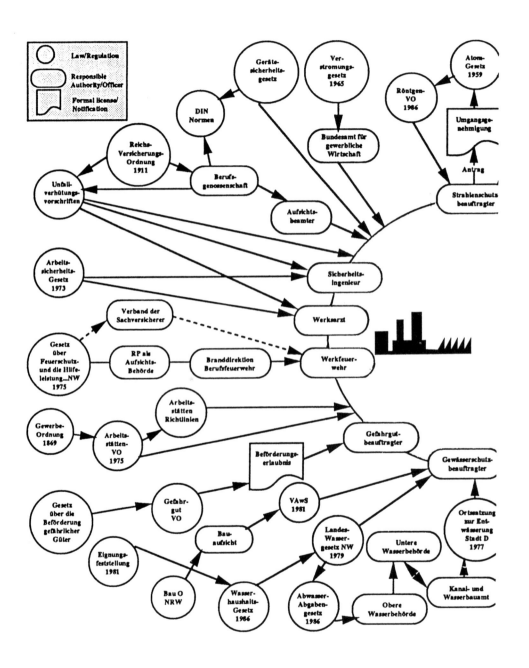

Figure 3.1   *Legal framework for chemical process plants in Germany*

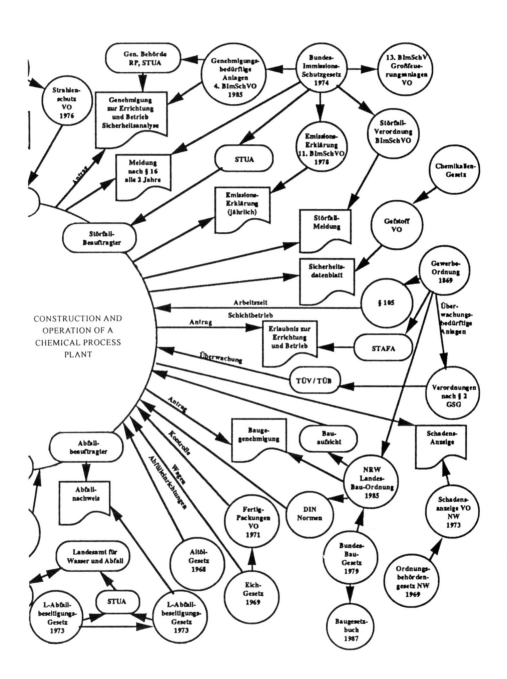

• constant inspection of the operation of chemical plants and their components by the operator and by so-called operator officers for special tasks;

• initial and periodical inspection of the plants and their components at set intervals by impartial experts and specialists;

• inspection by state authorities, initially during the authorization or licensing procedure, as well as in unannounced on-the-spot inspections.

3.1.1   Inspection by the operator

The operator is responsible for the safe operation of the plant. The obligation for constant monitoring according to the laws concerning occupational health, air pollution control, water protection, hazardous materials and the German National Insurance Code thus lies with the operator.

During normal operation the plant is constantly monitored and checked or inspected recurrently by different persons or service departments from various viewpoints, according to a defined distribution of tasks (Figure 3.2).

The respective plant managers are responsible for the safety of their plants. Central service departments for plant safety, occupational health, fire protection and environmental protection are set up to support them.

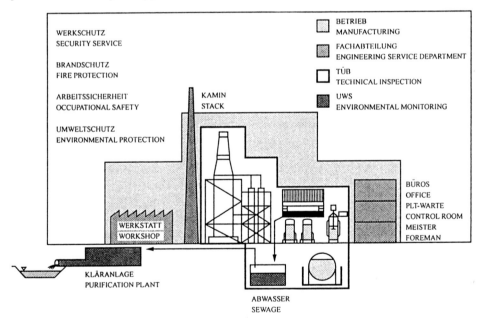

Figure 3.2   *Technical plant inspection at Bayer AG*

The following inspection activities take place:

• The plant operator monitors that operations proceed as intended — for example, with the help of the operating data displayed and recorded in the control room, and also by making inspection rounds.

● The functions of process control systems relevant to safety, such as protective devices and process analysis equipment, are inspected on the basis of maintenance schedules in consultation with engineering service departments.

● The Fire Protection Department inspects the buildings for compliance with the fire protection regulations — for example, the existence and correct functioning of the preventive fire protection installations.

● The Occupational Safety Department checks compliance with the regulations concerning hazardous materials, the regulations of the German Workplace Ordinance, and the Accident Prevention Regulations.

● The Environmental Protection Department inspects the outlets into the receiving water and/ or the individual waste water flows, and also the air purity in the works and the surrounding area.

● The Security Department monitors site traffic as well as facilities in particular need of safeguarding. General access checks reduce the risk of intervention by unauthorized persons.

● The Technical Inspection Department carries out all legally prescribed tests and inspections on plant components requiring recognized expert inspection or special inspection orders given by the operator. It represents a special feature of Bayer's safety organization. Officially recognized experts are brought together in this department in order to examine the suitability and correct functioning of safety-relevant plant components.

Any findings during these different inspections are documented and appropriate schedules set for corrective measures. Their definition and implementation lies within the responsibility of the respective plant management. After completion the service department concerned is informed.

Occupational health, fire protection, air pollution control and water protection are thus equally guaranteed by this comprehensive inspection system.

Maintenance

Necessary maintenance measures and alterations within the plants must be carried out with professional competence and great care. Respective responsibilities and procedures have been defined in a guideline 'Technical Shift Log, Work Permit and Process Modifications'. For technical work on the plant the organizational regulations serve to:

● realize comprehensible engineering planning before placing orders;
● guarantee a process-oriented distribution of tasks;
● ensure a uniform and safe implementation procedure;
● achieve uniform and comprehensible documentation of the plant's condition.

According to this guideline, the production, supply and disposal plants, warehousing sectors and pilot plants must keep a Technical Shift Log up to date which documents disturbances and defects of components and equipment and their maintenance. The work to be carried out and the procedures for handling orders are recorded in this log.

After the maintenance measures are concluded, their implementation and the achievement of the original plant condition (nominal condition) is confirmed. Damage and defects which have not been dealt with are transferred to the documentation of the following shift and handled accordingly.

Operator officers

Individual statutory regulations oblige the operator to appoint so-called officers for certain aspects of the operator's sphere of responsibility. The following officers are distinguished:

- Occupational Safety Officers;
- Plant Officers for Pollution Control (Environmental Protection Officers);
- Hazardous Incident Officers;
- Experts according to the Pressure Vessels Ordinance;
- Experts according to the Ordinance on Electrical Installations in Explosive Atmosphere;
- Hazardous Goods Officers;
- Water Protection Officers;
- Radiation Protection Officers.

These officers are assigned to check the respective areas of safety and health protection. Their tasks include monitoring of compliance with regulations, stipulations and requirements, and also regular inspection of chemical plants, paying particular attention to the correct functioning of technical and personal protection devices. The Hazardous Incident Officer is of particular importance in this context, with a duty to carry out physical inspections of every individual facility at regular intervals.

The officers have not only the right, but also the duty to report immediately any deficiencies they detect and to propose measures for eliminating them. They are not told how they should do their respective tasks; each sets up an action plan for the measures and completion is reported.

3.1.2   Inspection by independent experts and officially recognized specialists

Numerous legal standards require that impartial experts carry out safety engineering inspections of the plants (Table 3.2). For example, the Act on Equipment Safety stipulates that the inspection of installations requiring monitoring, such as pressure vessel plants, pipelines, electric installations in explosive atmosphere and specific warehousing facilities, be carried out by officially recognized experts.

The Ordinances and Technical Regulations allocated to this Act give concrete directions regarding the scope, depth and aim of inspections. According to these specifications, inspections of installations, equipment, machines and apparatus requiring monitoring are to be carried out by impartial experts at different stages:

- in the course of construction and erection;
- before commissioning (acceptance tests);
- at recurrent intervals during operation.

Table 3.2   *Type and scope of technical inspections (excerpt)*

UNIT	SUBSTANCES	REGULATION	TEST CYCLE (YEARS)	
Pressure vessels	All	● Pressure Vessel Ordinance (DruckbehV)	2	outside inspection
			5	inner inspection
		● Technical Rule Pressure Vessels (TRB 514)	10	pressure test
Filling plants	Flammables	● Technical Rule Pressurized Gas (TRG 790)	4	
	Highly toxic gases Flammable liquids	● Technical Rule Flammable Liquids (TRbF 111)	5	
Pipelines	Flammables, toxics, caustics	● Pressure Vessel Ordinance (DruckbehV) ● Technical Rule Piping Systems (TRR 514)	5	
Tank plants and pipings	Flammable liquids	● Flammable Liquids Ordinance (VbF)	5	
			3	(electrical equipment)
		● Technical Rule Flammable Liquids (TRbF 620)		
Tank plants Filling plants	Water pollutants	● Ordinance on Storage, Filling and Handling of water-polluting Substances (VAwS) ● Technical Rule Flammable Liquids (TRbF 620)	5	(electrical equipment only at commissioning)
Electrical equipment	Flammable liquids and gases	● Ordinance on Electrical Equipment in Explosive Atmosphere (ElExV) ● Directive on Explosion Protection (Ex-RL)	3	
Rail tankers		Hazardous Goods Transportation Ordinance		
Road tankers	as per Hazardous Substances Ordinance	● Railway (GGVE/RID) (Railway Tankers)	4	inner inspection
			8	pressure test
		● Road (GGVS/ADR) (Road Tankers)	3	inner inspection
			6	pressure test
Transportable containers		● Sea (GGVE/GGVSee) (all containers)	2, 5	inner inspection
			5	pressure test
Lifting equipment	All	Accident Prevention Rules of Mutual Industrial Insurers Association (UVV'en)	1	
Elevators	All	Elevator Ordinance (AufzV)	1	
Working machinery eg; centrifuges	All	Accident Prevention Rules of Mutual Industrial Insurers Association (UVV'en)	1	
			3	dismounted

In a major company such as Bayer AG, these inspection tasks are carried out by recognized experts from the central Technical Inspection Department. These experts have equal standing with the experts of the outside Technical Control Boards (*Technische Überwachungsvereine — TÜV*) who otherwise tackle such tasks. An internal company handbook outlines the type and extent of the inspections as well as the organizational procedures, and the instructions are binding on all involved.

Impartial experts regularly carry out inspections of installations and components at fixed intervals on the basis of water protection legislation, the law concerning the transportation of hazardous goods and various accident prevention regulations.

In the course of the recurrent inspections, a particular check is made on whether the statutory prescribed inspections and the inspection measures defined by internal agreements — especially functional tests — have been carried out by the service departments responsible. The expert inspects the maintenance logs and daily plant logs for this purpose. If weak points are discovered the expert, plant personnel and service departments together draw up measures, and an associated schedule, for eliminating them. The expert is informed when the resulting work is finished.

3.1.3   Inspection by state authorities

Even before a chemical plant is put into operation, it undergoes various safety inspection procedures by different specialist authorities — for example, building inspection authorities, factory inspectorates, industrial safety authorities and subordinate water protection authorities. The competent licensing authority is informed of the respective inspection result and issues the licence, taking the statements of the specialist authorities into account.

After the plants have been put into operation, they can be inspected and checked at any time, either on-the-spot or through the evaluation of requested documents by the regulatory authorities — these usually being the specialist authorities already referred to. These authorities may also commission impartial experts for this purpose.

3.2   *An integrated modular audit system*
K J Niemitz, J Köhler, C Jochum and N Schadler, Hoechst

The complexity of regulations and the diversity of a chemical unit and associated organization define a system which has plenty of opportunities to fail. To reduce the number and extent of these failures is a continuous challenge for the chemical industry. It is certainly true that the implemented safety concepts do a good job in general. However, this is not 'bad news' and therefore difficult to communicate. It has to be admitted that there are still enough incidents worldwide to focus the attention of the authorities and the public on the efforts of the chemical industry to address this challenge.

One of the tools available to reduce the number of incidents is the audit. This involves systematically asking the right people appropriate questions, evaluating the answers, and defining an action plan (see Figure 3.3). Thus, an audit can be a solid basis for moving efficiently towards a better overall performance.

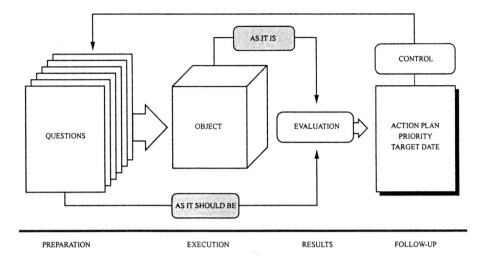

Figure 3.3 *The course of an audit*

In line with a proposal to the board of management of the Hoechst Corporation, an internal audit program was set up, aiming to increase the level of safety and to improve process technology. The audits within the program would have to result in prioritized lists of proposed investments, covering all aspects of the manufacturing process including process safety, environmental and health issues, process technology, logistics, training systems and the management organization.

For this reason Hoechst has decided to establish an integrated audit system to incorporate as many different audit aspects as possible under one roof. This section gives an overview of the system: how it was created, how it works and where the company wants to go in the near future.

3.2.1 Concept

An audit system is of course only one part of a safety management system. It produces a time- and place-dependent picture of something, and therefore cannot serve as a continuous tool to maintain a prescribed level of safety. The findings can be summarized and prioritized, and in this way serve as a foundation for a capital investment plan. After completion a higher level of performance will be reached, which is the main goal of any audit.

In the past a lot of different types of audits have been established. Normally they are carried out by different departments or external experts, coordinated more or less conveniently with respect to what is being audited. Information flow between the different audits is not established formally. The results are documented in different ways, and preparation for the audit is an additional effort for the facility staff over and above their normal activities. Therefore, it is not surprising that a plant manager generally does not look forward to the next audit. This attitude is, of course, totally unproductive for any audit because success is strongly dependent on the support from the audited unit. This applies not only to the execution of the audit but also in the follow-up phase.

The intention of the integrated audit system has to be considered alongside specific German regulations. According to the *Bundesimmissionsschutzgesetz* (German Federal Immission Protection Law), the authorities are allowed to force a plant manager to invite an inspection by an external expert. This expert has to inspect the plant according to a given plan and has to prepare a report containing the findings. These findings must be addressed by the plant manager within a prescribed time determined by the importance of the findings, the possibility of doing something about them and the costs. In the context of this regulation a programme was set up two years ago in the state of Hessen which covers all units falling under the Störfallverordnung (the German implementation of the Seveso Directive[1]). In addition, the state authorities initiated a special programme (ASCA), a typical compliance audit covering certain regulations with special emphasis on occupational safety.

Besides these spontaneous external activities there are so-called Beauftragter (commissioners) whose job is to assist units to comply with internal and external regulations (see Figure 3.4). Beyond that, they are also obliged to check the unit, which in fact is similar to auditing.

An integrated system should cover all the different sub-systems, internal as well as external, in order to coordinate the internal activities in an efficient way and to replace the current external programmes or inspections, which would be in full agreement with the Seveso II Directive[2].

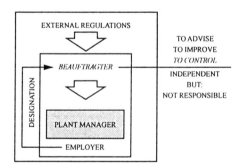

Figure 3.4   *The role of the* Beauftragter

3.2.2   Design

To add an internal audit programme to existing programmes is a big challenge and can only succeed if certain conditions are fulfilled:

● The system which replaces the internal programs already existing must be an integrated one.

● It should focus on safety-related issues as well as on process-related problems.

● Technical aspects are as important as organizational issues.

● Compliance with internal/external regulations should be part of the system.

A modular system was chosen to structure the contents of the audit lists. A total number of 18 modules was set up:

1. Validation of plant equipment documentation (for example, P&IDs).
2. Process hazard analysis.
3. Measures to detect deviations from normal operation.
4. Safety concept to minimize consequences of incidents.
5. Fire/explosion prevention.
6. Quality- and safety-relevant instrumentation.
7. Energy supply system.
8. Health issues.
9. Standard operating procedures.
10. Training/knowledge of workers.
11. Maintenance programmes.
12. Emergency response planning.
13. Management systems.
14. Hazardous materials.
15. Process and production improvement/optimization.
16. Environmental issues.
17. Work permits.
18. Occupational safety.

Each module consists of a standard procedure describing the audit process in detail, such as who is involved and which documentation is needed. The main part of each module is a questionnaire. As far as possible the questions have been formulated so that they can be answered 'yes' or 'no'. This allows analysis of the results which easily brings out findings which are not only relevant for certain units but appear to identify a problem of wider importance. Of course such simplicity is limited to certain questions, especially those regarding compliance issues. The more important questions regarding safety- or process-related problems can only be presented in open form, and require an expert to come up with the actions for further improvement in line with the specific conditions of the audited unit.

All modules contain the four elements of an audit (question, discussion, inspection, examination) to different degrees (see Figure 3.5, page 20). For example, 'Validation of the plant equipment documentation' (module 1) focuses mainly on examination of the documentation (P&IDs and hazard analyses) and inspection of the equivalent equipment in the plant, whereas 'Training/knowledge of workers' (module 10) is characterized by open questions and discussions with the facility staff.

The essential aim of the audits is to identify deviations, which can be classified according to the element they are focusing on (see Figure 3.6, page 20):

● Process — deviations between the actual risk related to a certain unwanted scenario and the generally accepted risk defined by internal and external standards. Example: the measures to avoid a runaway scenario are not sufficient to avoid the need for an additional blow-down system.

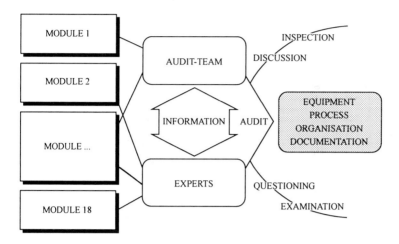

Figure 3.5   *The design of the integrated audit system*

● Organization — deviations between the documentation and what actually happens. Examples: the manager checks the plant regularly but not according to the prescribed procedure; plant personnel are not aware of the hazards of dangerous materials, although they are informed of them regularly by the plant manager.

● Equipment/documentation — deviations between prescribed standards and regulations (internal as well as external) and the actual implementation. Examples: according to the *Gefahrstoffverordnung* (Hazardous Materials Regulation) a form describing procedures to be followed while handling the material has to be available, but it is not; the P&IDs are not consistent with the actual layout of the plant.

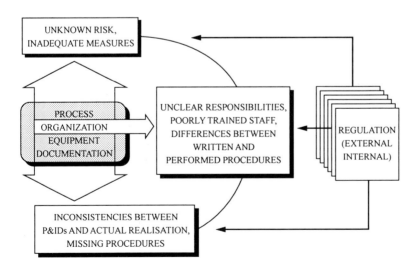

Figure 3.6   *The different types of deviations*

What all these deviations have in common is that they enhance the risk of incidents and accidents. They are linked together to different degrees. If no procedure is written to prescribe the way to update the P&IDs, it is likely that the P&IDs will not represent the actual situation in the plant. A wrong P&ID as a starting-point for a risk analysis leads to unreliable results.

The deviations regarding the equipment and documentation have a static aspect. Once they are recognized they can be corrected easily relatively quickly. Organizational deviations, on the other hand, can be characterised as dynamic problems, because the behaviour of the plant personnel has to be changed. This generally takes a lot of time before new habits become an obvious part of daily life. The third type of deviation results from a classical hazard analysis concentrating on critical parts of the whole process. The findings within the other modules are included to obtain additional information about the risks in certain areas. If, for instance, organizational measures are part of the safety concept and the interviews with the personnel have clearly demonstrated that they are not aware of their duties in an emergency, additional technical features to reduce the risk must be considered. If this is not feasible, a programme has to be set up to improve the training of the appropriate people.

Each deviation corresponds to a finding proposing an action to correct the deviation within a certain time. The time frame is determined by the seriousness of the deviation, the complexity of realization and the costs. Normally the time frame ranges from three months (if only formal aspects have to be corrected) to 18 months (if parts of the safety concept have to be reviewed, including the implementation).

The modules have been created by implementing checklists which already exist. This was done in such a way as to eliminate most of the redundancies. Nevertheless, a certain degree of redundancy was maintained as a basic principle, to guarantee that more than one look is taken at possible problem areas.

The modular structure was chosen for different reasons. In principle it enables specific, so-called target audits to be carried out. Extending the whole audit program to a longer time scale, a quasi-continuous internal self-controlling system can be realized. A further advantage is the minor impact on production. A second reason is driven by the demand that the contents of the modules have to be updated according to the findings of the audits, and to the changes of requirements of internal and external regulations. Therefore, each module is linked to an appropriate internal service unit. These units are responsible for maintaining the module, in order to ensure that extent and content are adequate, sensible and practical.

3.2.3   Practice
The audit is conducted by an audit team which consists of three people: two internal members, one more safety-oriented and the other more specialized in process aspects. The third position is filled by an external, officially acknowledged (and therefore independent) expert to ensure that the results of the audit will be approved by the authorities. It is hoped that in the near future this will only be necessary in special justified cases.

One part of the audit is carried out by the audit team but certain modules are handled by appropriate experts to ensure that highest competence is available; this is an important condition for the acceptance of the whole audit.

The integrated aspect of the audit is not created by adding all relevant modules to one audit system. It is mainly achieved by the coordinated flow of information between the different modules. This is one way to increase the chance of discovering underlying problems. The flow of information has been fixed in the working procedure of each module. Of course this prescribed information flow does not stop an auditor passing important information gathered during the audit of one module to other modules.

The extent of the audit system requires a detailed schedule. This is prepared by the audit team, to ensure that the modules are handled in a sensible way. If this is not possible, there is a chance that additional meetings will be needed to address unanswered questions.

3.2.4   Results
The whole project was started at the end of 1995. In March 1996, two units served as pilot plants to test the tool and to identify its weaknesses. The system was corrected and completed. Currently, the program is covering all units of the speciality chemicals division at the Hoechst site in Frankfurt.

The experience to date is varied and can be summarized as follows:

● The modular concept proved able to handle the problems posed by the complexity of a chemical unit.
● The exchange of information between different modules is important to identify the whole spectrum of possible deviations.
● The developed checklists are an adequate tool to cover a sufficiently wide range without limiting the auditor with respect to the overall targets.
● The resulting paperwork is limited to completed checklists and the summarized findings; no further paper is produced.

The impact of the audit on the workload of the units is not negligible, of course, because it takes place in addition to all the other activities. Nevertheless, the basic principles of the procedure have been accepted by the audited units.

3.2.5   Future developments
The Integrated Audit System will become an important part of an efficient safety management system. It describes the extent of the internal inspection activity. This corresponds to the role of the internal service units in supporting the production units to reach their objectives. Therefore, this system is primarily a tool for the service units. A database was designed to manage the questionnaire, which is an important condition for a structured and easy way to analyze the results.

Any audit system is an important part of an internal safety management system. To maintain a high level of competence and efficiency, a certain degree of manpower has to be provided. This internal effort should be acknowledged by the authorities: associated with spending the internal effort is the wish for deregulation, perhaps as a specific German problem. This wish is in full accordance with the Seveso II Directive[2]. External inspections should be reduced to a formal system-oriented audit, which only checks whether the implemented safety management system is working properly.

From this point of view it is not of vital importance to extend the main intention of the audit system towards a tool to quantify the performance of a unit in terms of percentage numbers. The comparability of these numbers exists only within the same unit at different times the same audit is executed. This could be an indicator to verify whether the findings regarding systematic errors have been corrected in an efficient way.

Nevertheless, the main goal of the audit is not to quantify the results. Hoechst is strongly convinced that the higher level of performance reached after all findings have been corrected must lead more or less automatically to a better overall performance in terms of lower incident rates or improved production results.

3.2.6   References

1.   *European Council Directive on the major accident hazards of certain industrial activities*, Directive 82/501/EEC, 24 June 1982, (European Community, Brussels, Belgium).

2.   *Common Position (EC) No 16/96 on Council Directive 96/.../EU on the control of major accident hazards involving dangerous substances*, 19 March 1996 (Council of the European Union, Brussels, Belgium).

3.3   *Technical audit/inspection and qualitative safety rating*
G Burgbacher, TÜV Südwest

The Seveso II Directive requires, in addition to the technical and organizational objectives, the implementation of a safety management system[1]. Therefore, methods are needed to enable operators and authorities to verify the safety management system, in particular the system elements, and to validate the system's efficiency, effectiveness and reliability.

The Directive defines two levels of requirements for safety management systems. There is a general requirement for all companies to produce a major accident prevention policy, designed to guarantee a high level of protection for human beings and the environment with appropriate means, including management systems.

In addition, the operator of a site covered by Article 9 of the Directive (corresponding to a larger inventory of hazardous substances) is also required to demonstrate in a safety report (in German it is called a 'safety analysis') that both the major accident prevention policy and the safety management system for implementing it have been put into effect in accordance with the information in Annex III of the Directive.

A general overview of the protective objectives and strategies is provided in Figure 3.7, page 25. It explains the change from Germany's rule-based inspection, with the associated component-related approach, to the overall system supervision which is characteristic of the Seveso II Directive.

Typical examples of safety management practices in industry are described in reference 2. Experience with several companies shows that system elements commonly implemented correspond with the requirements of the Seveso II Directive (Table 3.3).

3.3.1   Audit program design

Whereas the individual system elements can normally be identified and evaluated, the interlinking of the procedures, rules and other management tools to form a functioning safety management system is currently in its infancy. Auditing in this phase of safety management system implementation is very difficult, since no real yardstick yet exists. Therefore, it must always be decided what type of audit should (or can) be conducted, for example:

● on-line audit/monitoring;
● technical audit/inspection;
● compliance/legal review/systems audit;
● organizational/total quality management (TQM) audit.

This important decision can be clearly illustrated with an example. Here are two types of review:

1.   The piping and equipment for a process can be inspected, and the absence of a required pressure relief device in a technical system can be identified, and

2.   It can be ensured with the implemented safety management system that, for example, pressure relief valves have been designed, installed, operated and maintained in accordance with company standards.

The first of these reviews addresses a particular hazard found at a specific time. It could lead to the correction of the hazardous condition without addressing its underlying cause. The second addresses the management system which is in place to prevent such hazards from arising. Detection of a deficiency in the system could result in the prevention of hazards in the future.

Table 3.3   *System elements required by the Seveso II Directive*

Accident prevention policy	Periodic review
Organization	Systems of work
Communications	Mechanical integrity, inventory
Staff selection and training	Emergency plans
Hazard identification and assessment	Improvement
Documentation, safety report	Investigation and reporting
Prevention, mitigation	Contractors
Modification control	

This means that the objective of auditing is specific, depending on whether or not a safety management system is already implemented. The following questions must be answered before proceeding[3]:

● Is the audit to concern the entire company or only parts of it?

● Is the audit to be system-oriented and is it to prove that a functioning safety management system is implemented in accordance with the relevant company guidelines and legal regulations (system audit)?

● Is a certain procedure to be examined for susceptibility to failure and for effective corrective measures being available (process audit)?

● Is the quality of the safety measures to be evaluated (performance measurement)?

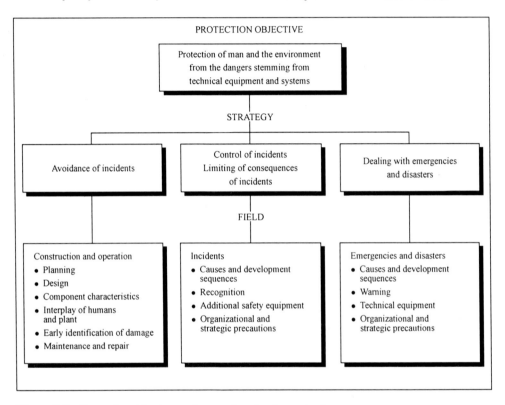

Figure 3.7 *Protective objectives and strategies of safety technology*

All parties concerned must be obligated to adhere to the objectives, since only then can all necessary information be made available for evaluating the 'object of the audit'. Eventually, the 'design' of an audit programme requires certain points to be taken into consideration, such as the scope, frequency and intensity of the audit, the qualifications (experience, knowledge, neutrality) of the team personnel, report preparation, quality assurance and the handling of follow-up measures. Because there is no ideal solution here, it is imperative that clear company goals be defined and a consistent procedure be decided on, prior to the start of the audit. To do this, it is also necessary to take the complexity of the system into account (Figure 3.8, see page 26).

3.3.2   Technical audit and inspection practice

Current practice for conducting safety audits tends to be more one of a combined procedure, particularly in the case of medium-sized companies and in some cases also for plants within large companies. Technical audit/inspection and legal review are combined in such a way that the operator and government authorities obtain documents and results with the audit report which not only reveal the current state of plant safety, but also describe the measures which will be taken to correct and improve the safety management system.

Technical inspection/audit and the legal review should be structured as business processes and associated with the life-cycle. Table 3.4 shows an example for inspection after major changes, which could be elaborated analogously for inspection at other stages of the life-cycle such as during or after commissioning. This procedure has the advantage that both corporate and government needs and requirements are met, and that redundant work is generally avoided. Since in Germany separate safety inspections are included in this procedure as well, the combination also meets the requirements concerned with comprehensive plant monitoring.

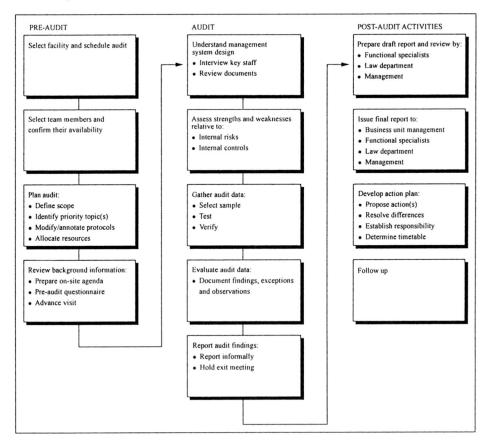

Figure 3.8   *Typical steps in the safety management audit process*

Table 3.4 *Inspection time: inspection after major changes (taken from reference 4)*

INSPECTION GOAL	BASIS FOR INSPECTION	INSPECTION PROCEDURE	INSPECTION CERTIFICATION
Determination of the conformity between system, system operation and certification (notices of approval and application documents†) and the major changes	● Notices of approval and application documentation* ● Safety analysis (as per Section 7 of German Accident Ordinance) or safety considerations for certification of the basic obligations being met (as per Sections 3 to 6 of German Accident Ordinance) ● Expertises/Position papers ● Operating manual including the operating instructions ● Documents for employee training ● Authorizations (as per Sections 55, 58a of German Federal Immission Protection Law) ● Maintenance documentation ● Alarm and danger prevention plan ● Documents from other legal areas ● Inspection documents of inspections already conducted	● **General inspection** Inspection of whether the safety analysis is updated in accordance with Section 8 of German Accident Ordinance ● **Documentation inspection** Inspection of the system documentation with the following inspection procedures: - Agreement of materials, processes and system with documentation - Agreement of safety concept and safety-relevant process variables and their limits for alarms and actions with the documentation - Realization of the technical measures - Implementation of the organizational measures in operating instructions/ operating documentation ● **Operating inspection** Inspection as formal inspection, or independent expert inspection by inspection personnel as per Section 29a of German Federal Immission Protection Law, or operator inspection	● Expert report ● Inspection record ● Expertises ● etc

† Application documentation as per Section 4 of 9th German Federal Immission Protection Ordinance

* Particularly piping and instrumentation flow charts, equipment lists, set-up diagrams, material data sheets, process descriptions etc.

By letting, for example, authorized TÜV experts (section 29a of the *Bundes-immissionsschutzgesetz*, the German Federal Immission Protection Law) participate in the audit process, it is also ensured that experts are available to the company who are familiar with the technology, organization and safety culture.

Audit techniques

A tour of the plant before the start of auditing particularly supports the conception and development of the auditing framework (scoping). This purposeful procedure generally allows a comprehensive audit plan to be adopted. This audit plan addresses the following questions:

● Which audit steps must be carried out?
● How comprehensive should these steps be?
● Who does what and in what order?

Part of this planning process also includes the specification of a structured, representative cross-section of company employees to be interviewed. Interview appointments must be announced well in advance to everyone involved.

It is advisable to begin the actual audit activities with an opening meeting on site. This enables the audit team to present the goals and to make the integral approach of auditing clear to the plant management. This type of on-site audit involves the following five basic steps:

● Understanding the management system.
● Estimation of strengths and weaknesses.
● Collection of audit data.
● Evaluation of audit data.
● Report preparation and explanation of audit findings.

Audit tools and methods

Audits are generally supported with suitable computer-aided tools, such as collection records or checklists, questionnaires and current specifications. The primary methodological tool is the interview technique.

Collection records are written lists which support the step-by-step procedure for the collection of data, and are generally used where a certain standardization from audit to audit is required. They are used both as an annex to the audit plan (completion of the records) and as proof of quality for the audit team within the specified audit procedure.

In addition to these collection records of basic information, suitable checklists are used as questionnaires — particularly during the interviews.

Regardless of the respective backgrounds of the interviewees, the duration of the interviews and the number of forms used, audit interviews are generally conducted in the following sequence:

● Planning.
● Opening, conducting, conclusion.
● Documentation.

It is important not to perform the interviews as routine conversations. The intention of the interview is to achieve interaction (dialogue, open exchange of opinions) between the interviewer and the employee. In general, it is advantageous to include experts with practical interview experience in the team of auditors.

The auditors are responsible for guiding the conversation. They hold the guide with the audit question lists in their hands, ask the questions and change the topic when they have received 'sufficient' information on a particular issue. The auditors consider responses critically, and should aim for additional information or action from those being audited, as necessary. For example, they may require evidence for specific statements or demonstration that certain circumstances exist on site.

Especially in this phase, the purpose of the safety audit must be clear. It is not to prove that a particular person has made an error, but rather to eliminate the jointly determined error with suitable measures, and thus to improve the safety of the plants.

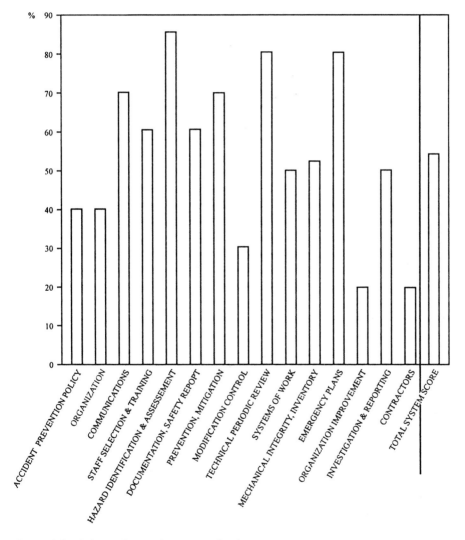

Figure 3.9 *Safety rating results — example of system score*

3.3.3   Safety performance — results

Safety management system audits performed in the initial implementing phase cannot be expected to attain 100 % compliance. According to the definition, auditing is an adequate procedure to measure the performance and to verify the tasks'for ongoing improvements.

Figure 3.9, page 29, shows the results of an actual system audit which correspond with an average level of compliance of 55%[5]. The subsequent review confirms that improvement is necessary and defines all further steps for completion of the safety management system.

3.3.4   References

1. *Common Position (EC) No 16/96 on Council Directive 96/.../EU on the control of major accident hazards involving dangerous substances*, 19 March 1996 (Council of the European Union, Brussels, Belgium).

2. EPSC, 1994, *Safety Management Systems: Sharing Experiences in Process Safety* (Institution of Chemical Engineers, Rugby, UK).

3. Burgbacher G, *Sicherheitsaudit als Baustein effektiver Management-Systeme — Ziel, Zweck, Durchführung*, 1995, TÜV Südwest Seminar *Sicherheitsaudit und Sicherheitsmanagement in Chemieanlagen*, Mannheim.

4. *Abschlußbericht Arbeitskreis 'Anlagensicherheit'*, 1996, TAA-GS-11, Technischer Ausschuß für Anlagensicherheit beim Bundesminister für Umwelt, Naturschutz und Reaktorsicherheit, Köln.

5. *Abschlußbericht Sicherheitsaudit Chemieanlage*, 1995 (TÜV Südwest, Filderstadt).

3.4   Condition measurement
G Koppen and W N Top, DNV

Condition measurement involves at least two measurements: the measurement of physical conditions and the measurement of risk.

The measurement of physical conditions is done through visual observation using checklists describing the items to be observed. The observation provides a numerical value indicating the hardware conditions found to be in good order.

The measurement of risk follows the risk ranking method used with DNV's risk-based inspection activities. Risks are calculated per equipment item and expressed as a combination of damage area (consequence) and likelihood of occurrence. Cumulative figures can be arrived at to indicate the 'risk level' of a plant or a part thereof.

3.4.1   A conceptual view on condition measurement

Hardware conditions are the end results of many activities in the organization which are part of the management system. Such activities include:

● hazard analysis;
● design of workplace and installations;

● maintenance and inspection;
● management of change;
● review of legislation, industry codes and standards;
● employee training;
● management leadership and supervision;
● purchasing of goods and materials.

Condition measurement is a means to:

● uncover initial design oversights;
● find deviations from original design specifications which may have been introduced through modifications;
● establish conformance with codes, standards and industry practices.

The measurement of (hardware) conditions and risks is mainly focused on the determination of the effectiveness of management system activities. As these activities are directed at eliminating and controlling substandard hardware conditions and related risks (so that losses will be controlled at a desired level), the measurement of those conditions and their risks can be seen as an indicator for the effectiveness of the 'upstream' control system.

There is also a secondary purpose associated with the measurement of (hardware) conditions and risks, namely the determination of substandard conditions for correction. Although condition measurement is not really intended to be an 'inspection' in order to generate items for correction, it is only natural that at least part of the non-conformities noted during the evaluation are corrected.

In this section, two methods for condition measurement are discussed: evaluation of physical conditions and risk-based measurement (risk-based inspection).

3.4.2   Evaluation of physical conditions

The evaluation of physical conditions is based on the process of carrying out observations, which are directed at hardware conditions and intended to determine whether or not the items observed comply with established criteria. The result is a comparison between the total number of observations and the number found to be in compliance. The end result is a ratio indicating the effectiveness of the upstream control activities. Under ideal circumstances all items observed should be in compliance and a 100/100 ratio will exist.

In order to rank the items observed relative to each other, they are normally grouped together in categories and provided with a value factor depending on the possible consequences of non-compliance. This is indicated in Table 3.5. The 'risk score' is calculated per category, using the compliance ratio and the category value factor.

Measurement of physical conditions
Measurement in the context of physical conditions evaluation is done using a form such as that shown in Table 3.5, page 32. On this form the items to be observed are listed, whereas 'what to look for' is provided through more extensive checklists. Compliance requirements are included in those checklists and can be as detailed as deemed necessary. This type of

Table 3.5 *Form for physical conditions evaluation*

CATEGORY/ITEM	NUMBER OBSERVED	NUMBER NOT IN COMPLIANCE	VALUE FACTOR	RISK SCORE
A. *General workplace conditions*			10	
3. Platforms/scaffolding				
8. Ventilation				
C. *Materials*			20	
12. Stacking and storage				
15. Waste disposal				
D. *Equipment*			25	
19. Lifting gear and equipment				
24. Hydraulic power systems				
F. *Emergency systems*			15	
34. Fire protection systems				
TOTAL			100	68

observation can be carried out by people with varying experience and qualifications, depending on the purpose of the evaluation as well as on the issues/items being observed. The frequency of these evaluations depends on the purpose. More general aspects may be evaluated more frequently (as part of regular management system effectiveness evaluation), whereas detailed compliance evaluations (comparing conditions with codes, standards, etc) may be done less often.

Analysis of physical conditions measurement results
The results of the condition measurements can be used as follows:

● as a measurement of the upstream (management system) activity effectiveness;
● over a period of time to compare an entire site or units on the site with itself or with each other. Progress as well as deterioration can be measured;
● as an indication of the level of compliance with codes, standards, etc, depending on the purpose for which the evaluation was made.

Follow-up on physical conditions measurement
The results of an evaluation of physical conditions may be seen as an 'end-of-pipe' indicator. Depending on the levels set by management, the results may trigger further analysis of the causes of deterioration and initiate further upstream problem solving and action. The results of the evaluation can also lead to corrective actions.

3.4.3 Risk-based inspection
The performance indicator here is based on the risk calculation and ranking method which is part of the risk-based inspection (RBI) approach of DNV. This approach uses risk assessment techniques in decision management concerning asset integrity. DNV has been active in this

field for several years and has developed a method for the American Petroleum Institute, shortly to be issued as the Recommended Practice on Risk-Based Inspection. DNV has also developed RBI software to assist in the process, and allow 'what-if' studies to be carried out. These studies can be carried out not only on existing plants (in order to optimize inspection and maintenance activities), but also for major modifications and when designing new plants. In the latter case the software is being used to arrive at a best 'risk picture', looking at a variety of mitigating measures.

The RBI process comprises an assessment of the hazards and risks involved with each piece of process equipment, leading to a specification of appropriate inspection methods and frequencies as well as other risk-reduction actions.

The risk calculation and ranking activities include the following factors:

- generic failure data of the equipment type under consideration;
- an equipment modification factor.

In addition to these two factors, a third aspect may be added:

- a management system modification factor.

RBI risk measurement

For the risk calculation and ranking, the likelihood and consequence are calculated and expressed in matrix form.

Assessment of the likelihood of failure is carried out by using a maximum of three aspects:

- generic failure data, obtained from data bases for equipment/installation items like centrifugal pumps, columns, filters, heat exchangers, piping, pressure vessels and storage tanks.
- an equipment modification factor in which aspects as listed in Table 3.6 are considered;
- a management system modification factor, based on API 750 and in which elements as listed in Table 3.7 (see page 34) are considered.

Table 3.6   RBI equipment modification factor: elements/aspects to be considered

TECHNICAL MODULE SUBFACTOR	MECHANICAL SUBFACTOR
Damage rate	Equipment complexity
Inspection effectiveness	Construction code
UNIVERSAL SUBFACTOR	Life cycle
Plant condition	Safety factors
Cold weather	Vibration monitoring
Seismic activity	**PROCESS SUBFACTOR**
	Continuity
	Stability
	Relief valves

The consequence of failure is determined from an estimation of the potential inventory release rate and the effects of mitigating systems. In principle, four consequence areas may be considered: flammability/explosion, toxicity, environmental damage and business interruption. From these, a consequence category is determined per equipment type or item and ranked from low (A) to high (E), as indicated in Figure 3.10.

Table 3.7   *RBI management system modification factor: elements to be considered*

1.	Leadership and administration
2.	Process safety information
3.	Process hazard analysis
4.	Management of change
5.	Operating procedures
6.	Safe work practices
7.	Training
8.	Mechanical integrity
9.	Pre-startup safety review
10.	Emergency response
11.	Incident investigation
12.	Contractors
13.	Assessments

On the basis of the likelihood and consequence factors, the risk level for each equipment item can be determined from the matrix and items can be ranked according to their risk. Using this method, inspection extent and frequency can be optimized by directing attention to those items requiring it, and inspection techniques may be selected on the basis of the damage mechanism established. This usually results in reduced inspection time and increased inspection intervals, which leads to increased plant availability.

Plant safety and risk management can be improved by using the results of the risk assessment and ranking method. The risk assessment per equipment item may reveal those items requiring attention, whereas other, low-risk items may be 'ignored'.

Analysis of RBI risk measurement results
The matrix presented in Figure 3.11 provides the results of an RBI study in a petro-chemical unit with more than 450 equipment items. The matrix is a risk indicator: the greater the number of items found in the upper right-hand corner, the greater the risk within the installation.

The results of the risk calculation and ranking can be used:

● as an indicator of the risk level of the installation;
● to establish the effects of risk mitigation measures;
● to compare units and processes on the basis of risk;
● to trend the risk development of a unit over time and during its life cycle.

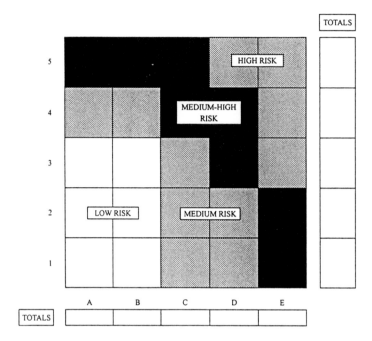

Figure 3.10 *Risk matrix*

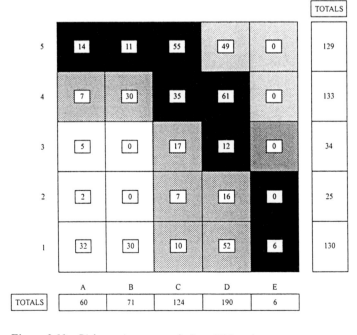

Figure 3.11 *Risk matrix — example from RBI study*

Follow-up on RBI risk measurement

The follow-up on the risk calculation and ranking (with the risk matrix as a 'performance indicator') obviously depends on the results as represented in the matrix. Depending on the risk level or distribution set by management, actions may be taken to reduce risk further.

Given a favourable picture, the RBI risk calculation and ranking may result in optimization of inspection and maintenance activities. Thus, attention may be given to equipment in need of it, while equipment with less risk gets less attention.

The risk of items with a high consequence factor and a low likelihood rating cannot be decreased by management systems but only by technical solutions — for example, by reducing the maximum inventory associated with one single loss-of-containment incident.

The risk matrix and related software allows further analysis to estimate risk development over time — for example, when nothing is done and the installation is run without any particular inspection or maintenance activity. At the same time the influence on risk of introducing certain risk-mitigating measures can be analyzed.

Besides providing a 'risk indicator', DNV risk-based inspection offers an auditable and transparent method developed for the process industries. It can lead to a reduced equipment failure rate, reduced inspection time and increased plant availability, and thus result in increased profitability.

3.5   *Developing a safety performance measurement system*
J H Christiansen, Borealis

This section describes the work done in developing a performance measurement system for Borealis, which consists of a conglomerate of sites established through several acquisitions over several years. Although various aspects of safety performance are covered, the emphasis is on operational safety.

3.5.1   *Introduction*
In 1992 Neste and Statoil decided to investigate the potential for combining their resources in the polyolefins area, thereby creating a new, more viable force. The outcome was Borealis, which became operational on 1 March 1994 and is currently Europe's leading polyolefins company.

In addition to operating polyolefin plants, Borealis operates three crackers and four minor compounding plants. The Borealis Group also includes a manufacturing company, which produces automotive parts.

The major operations are located in Belgium, Finland, Norway, Portugal and Sweden. The plants are all of different age and origin, and had different cultures: different companies owned the plants at the time of acquisition, and various plants had been owned by other companies in the past.

When the company became operational, the first priority was to develop a health, safety and environment (HSE) policy, to be followed by corporate guidelines. In order to have a platform from which a common safety culture could be established, the focus was on two major elements:

● the attitude towards accidents and near misses: reporting and investigation, management commitment and activation of all employees;

● a common set of minimum HSE requirements and an HSE management system, in order to get all the sites to the same level of HSE performance;

3.5.2 Injury performance measurements

With respect to accidents and near misses, a reporting system was set up to record the performance at all sites. The definitions of the important parameters were taken from the European Chemical Industries Council (CEFIC) definitions in order to be able to report to CEFIC on the Group's performance.

The following parameters are measured and recorded monthly:

● lost-time accidents (LTA), for own employees and contractors;
● restricted work cases (RWC), for own employees only;
● medical treatment cases (MTC), for own employees only;
● total recordable injuries (TRI = LTA + RWC + MTC), for own employees only;
● near misses, for own employees and contractors;
● lost-time accident frequency: LTA per million hours worked;
● total recordable injury frequency: TRI per million hours worked;
● near-miss reported frequency: number of reported near misses per million hours worked.

The LTA frequencies are reported on a rolling 12-month basis — that is, the LTA frequency for the last 12 months.

TRI, the number of total recordable injuries, is considered a very important parameter. When looking at the background for injuries, it is obvious that the same mechanisms are the basis for all kinds of injuries, whether they result in lost time, restricted work or medical treatment.

The work and attention on injury reduction have shown good results with respect to reducing lost time accidents. The TRI is a parameter that is being followed more closely than other parameters as it is the most appropriate representation of how the sites are handling the injury reduction program.

3.5.3 Auditing HSE performance

In order to have some common set of requirements against which it is possible to measure the sites, the minimum HSE requirements were produced. These requirements are in accordance with good engineering and operating practice in industry, and reflect two basically different approaches. The first and general part of the requirements is related to systems that need to be in order if the site is to operate a safe plant. The other parts are more technical: detailed requirements relating to health and environment and to the safety aspects of the plants. Due to the complexity of the major production sites, the minimum requirements are geared towards these sites, while the minor sites have to assess the requirements and delete those which are not applicable.

In order to assess the site's performance, these requirements are controlled by audits. Corporate HSE staff set up an audit schedule that is reviewed annually by the Executive Board's HSE committee. The production sites are scheduled to be audited every three years. Separate health/environment and safety audits are carried out at the major production sites, while combined HSE audits are carried out at the minor sites (smaller compounding sites).

The main purpose of the audit is to assess the site's compliance with or deviations from the HSE management system and the minimum HSE requirements.

The audit teams are nominated by the corporate vice-president HSE&Q, and will normally consist of 2-5 people drawn from corporate HSE staff and from the local sites. The members will either have experience in the appropriate HSE field or be experienced senior line managers. At least two of the team should either have experience as a team member from several audits or have training as an auditor.

The audits are carried out as a combination of interviews and verification of findings, by checking documentation related to the HSE management system and the minimum HSE requirements. The interviews start with the site manager and carry on throughout the organization. The following are examples of documentation which is necessary for the audit: former audit results, organization handbook, training programmes and documentation of training, engineering handbook governing the engineering at the site, maintenance handbook regulating the maintenance, and HSE handbooks.

To help in the audit a set of performance guides has been issued. These guides describe each element in the minimum HSE requirements at four levels — poor, fair, good, excellent — with the possibility of giving ratings from 0 to 10 for each element. However, it is not recommended to publish an average value for the audit; the inclination to beat the values can easily become more important than the real value behind the audit — that is, to improve the HSE performance.

After the audit is finished, the team writes an audit report which summarizes the major findings:

- strengths: positive findings;
- weaknesses: areas that can be improved;
- recommended actions.

Detailed findings are included referring to the minimum HSE requirements and the HSE management system. The report also includes the rating for each minimum requirement element, according to the HSE audit performance guides.

The responsibility for corrective actions and improving the site's HSE performance remains with the site and the site manager. Within two months of receiving the audit report, the site must respond in writing with recommendations for actions to address the major findings. This action plan must be forwarded to the Executive Board, with a copy to corporate HSE&Q staff.

3.5.4   Operational safety survey

As the major operating sites have complex plants, it was decided to develop a system for surveying the safety and operation of these plants. This work is presently in its infancy; the intention is to survey two high-pressure polyethylene plants by the end of 1996.

The objective of an operational safety survey is to identify existing and potential process hazards which could result in explosions, fires, releases of toxic and flammable materials, or serious injury.

An operational safety survey has to be based on a set of operational safety standards. As Borealis has many plants, with different licences, one common set of operational safety standards has not yet been produced. In order to carry through the planned operational safety surveys and to assess the use of such surveys, the operational safety standards for one of the licensed plants were used as the basis for the surveys. The results of the surveys will then be used when writing the Borealis operational safety standards for high pressure polyethylene plants. It is the intention to try and develop similar standards for polypropylene plants and for ethylene crackers.

An operational safety survey consists of an intensive and systematic examination of the specified process or operation for hazards to personnel, property and the environment from both a theoretical and practical standpoint. It includes the process design, physical facilities and procedures. Emphasis is on factors not readily apparent through purely visual observations, such as reaction limitations, effects of impurities in process streams, suitability of fire and explosion control devices, equipment design, materials of construction and operating procedures.

For the high pressure polyethylene plants, the survey is carried out by a team nominated by corporate HSE staff consisting of a high pressure specialist, an experienced process engineer and a senior plant manager with experience of high pressure plants. The survey normally starts with a discussion with plant personnel and developing an interview list. This list includes plant management, shift supervisors, operators, maintenance engineers, engineering people responsible for design changes, inspection people and high-pressure specialists.

An important part of the documentation needed concerns operating instructions, updated P&IDs, maintenance documents, results of hazard assessments (including potential Hazops) and inspection programs.

The team looks at the design, flowsheets and operating procedures, and examines the plant extensively in the field, looking for areas that do not conform to good engineering and operating practice. This implies that those participating in the survey must have a high degree of competence within their area coupled with long experience.

The survey is detailed, and the examination of the plant equipment and instrumentation loops is a vital part of it. To help the team, a checklist of process hazards has been developed (see Table 3.8). Table 3.9, page 42, shows examples of some of the standards which are used to measure against. Examples of areas found are:

- possibility of overfilling tank, due to lack of high high level alarm;
- possibility of overfilling tank, due to automatic cutoff having been cancelled when modification was carried out;
- safety-critical interlock system that should have been in operation was not functioning;
- on-line vibration monitoring should be introduced on equipment that is inaccessible during operation;
- changes in operating conditions must be documented;
- review critical operating valves and mark them for easy operation;
- set up periodic inspection of supports and anchors on high pressure piping, including vibration measurements where needed;
- hands-on emergency training, operational emergency training.

Equal to the audits, the operational safety survey is rounded off with a feedback meeting, in which positive and major findings are presented. A report is then written, referring to the elements of the operational safety standard, showing findings and recommendations. A rating is not given in the survey, and the report's recommendation is meant to be a help for the plant in achieving continual improvement in its safe operation and maintenance.

The responsibility for corrective actions and improvement remains with the site management. A plan for corrective action is expected from the plant within two months of receiving the report.

3.5.5   Conclusion

The monthly reporting on injuries and near misses focuses on past performance and is more like steering the boat by looking at the wake rather then looking ahead.

By auditing the HSE management system and the work being done at the sites against the minimum requirements, there is an opportunity to look ahead and set the course. Such auditing in Borealis is in its second year, and the experience with it shows improvement at the sites as a result of the audits. It also shows that the sites appreciate the audits and a good cross-fertilization between sites is developed by selecting the audit teams carefully.

Further, the experience with operational safety surveys for high-pressure polyethylene plants shows good results, and an extension to other plants will be discussed.

Continuous improvement is part of the Borealis Group's policy. It must therefore be understood that even though this section describes the way in which the company presently measures performance, the techniques will be monitored and possibly changed as more experience is gathered.

Table 3.8 *Process hazard checklist*

CATEGORY	SUBJECTS TO BE INVESTIGATED
Storage tanks	Design, separation, inerting
Dikes	Capacity, drainage
Emergency valves	Remote control, hazardous materials
Inspections	Flash arresters, relief devices, vents
Procedures	Contamination prevention, analysis
Specifications	Chemical, physical, quality, stability
Limitation	Temperature, time, quality
Pumps	Relief, reverse rotation, identification
Ducts	Explosion relief, fire protection, support
Conveyors, Mills	Stop devices, coasting, guards
Procedures	Spills, leaks, decontamination
Piping	Rating, codes, cross-connections, leaks
Procedures	Start-up, normal, shutdown, emergency batch sequence checklist
Loss of utilities	Electrical, heating, coolant, air, inerts, agitation
Vessels	Design, materials, codes, access
Identification	Vessels, piping, switches, valves
Relief devices	Reactors, exchangers, glassware
Review of incidents	Plant, company, industry
Inspections, tests	Vessels, relief devices, corrosion
Electrical	Area classification, conformance, purging
Operating ranges	Temperature, pressure, flows, ratios, concentrations, densities, levels, time, sequence, records of critical process variables
Ignition source	Peroxides, acetylides, friction, static electricity, heaters
Compatibility	Heating media, lubricants, flushes, packing
Controls	Ranges, redundancy warranted, fail-safe, calibration, inspection frequency, adequacy, backup during repairs
Alarms	Adequacy, limits, fire
Interlocks	Tests, by-pass procedures
Relief devices	Adequacy, vent sizes, discharge, drain
Process isolation	Block valves, fire-safe valves, purging
Instruments	Air quality, time lag, reset wind-up
Hazards	Fires, runaways, vapour clouds
Computer	Malfunction effects, protection
Ditches	Flame traps, reactions, exposure, solids
Vents	Discharge, radiation, mists, location
Sampling points	Accessibility, ventilation, valving
Fixed active protection	Sprinklers, deluge, monitors
Passive protection	Lagging, insulation, materials of construction
Flammable gas detectors	Reliability, applications
Fire walls	Adequacy, condition, doors, ducts
Drainage	Slope, drain rate

Table 3.9   *Example of checklist for process standards*

PROCESS MATERIALS

● Explosive concentrations of unstable compounds

● Relative feed rates of two or more reactive materials

● Safeguards, double valves and vents, automatic valve interlocks, disconnection of lines, avoidance of fixed piping cross connection, slip blanking

● Provision to prevent weather conditions from creating hazards

● Different materials stored in multi-compartment tanks

● Piping systems for heat sensitive materials, insulation, electrically bonded

● Venting of storage tanks containing flammables

● Piping and valving is arranged so that polymerisable material which can readily decompose, does not build pressure, and cannot be trapped or be stagnant

---

PROCESS FACILITIES

● Pressure relief devices on process vessels, devices in fouling services at more frequent intervals

● Vent pipe and stacks from pressure relief devices and hand vents, designed and installed to minimise accumulation of hazardous vapours in buildings and at ground level

● Process vessels are tested and inspected in accordance with guidelines and public requirements

● Other process equipment, maintenance equipment and miscellaneous equipment and devices are checked with such frequency and methods as will assure their continual safe operations and use

● Air intake for process use, for building or room pressurisation or for instrument purging are located in an acceptably safe position, and combustible gas detectors and alarms are installed in the inlets

● Piping is properly supported and free of excessive vibration

● Equipment and piping identified by numbers, name, or colour coding

● Utility stations have different couplings for steam, air, nitrogen and water to prevent mix-up

---

PROCESS CONTROL

● Instrumentation is adequate for control of critical process parameters, and redundancy is provided where failure of critical control instrument would create a hazardous condition. The redundant instrument is truly redundant

● All control valves are designed to move to a safe position in the event of instrument air failure, and/or instrument electrical power failure, or actuation of the emergency shutdown system

● Critical controls can be manually by-passed either by use of motor valve by-passes or control by-pass system while the equipment is in service

● Adequate safety margins exist in the means and methods for control of critical process parameters

---

4 Measuring systems and procedures: practical examples

The second area of safety management inputs is concerned with systems and procedures to operate and maintain the plant and equipment in a satisfactory manner and to manage all associated activities (see Table 4.1). Monitoring activities in this area include compliance audits, systems integrity audits, and overall management audits.

Table 4.1 *Safety performance measurement — monitoring the inputs*

	Regular inspec-tions and audits by local staff	Periodic in-depth inspections, assessments, audits of specific aspects	Overview assessments and audits by independent experienced assessors
Plant and equipment			
SYSTEMS AND PROCEDURES			
People			

This chapter presents four examples of monitoring activities in the systems and procedures area. To begin with (section 4.1), monitoring the implementation of local procedures at ICI is described. This involves regular systematic checks of an activity against local procedures and site instructions, which in ICI is known as 'operational auditing'. The associated audit process is described, and in particular the annual plan, the audit checklist which is included in each procedure, and the reporting phase are discussed. The section contains examples of an annual plan, an audit checklist, and a report form.

Next, a contribution from Dow (section 4.2) is concerned with measurement of safety performance within Responsible Care. A system of self-audit is described, which is called 'operating discipline'. This covers 13 areas, and one of these areas, management of change, is discussed as an example. Six key elements are addressed in evaluating the management of change process. The questions which are to be answered for these elements in plant self-evaluation of management of change are presented, and also the associated scoring system. The section concludes with discussing the validation stage.

The contribution from Exxon (section 4.3) deals with performance measurement of their safety or overall risk management systems. After introducing the company's operations integrity management system, the paper focuses mainly on one of the assessment steps: the self-assessment of the system. Two important dimensions are discussed which are considered necessary for a good management system and thus should be reviewed in evaluating the system: status and effectiveness. In conclusion, the overall system evaluation process is described, in which system design, implementation status and effectiveness are reviewed.

The final section of this chapter (section 4.4) is concerned with measurement through auditing and assurance at BP. Its integrated HSE management system is introduced, in which policy, expectations and assurance are key aspects. The auditing process used at the workplace to review and measure HSE activity and an associated upward assurance process for management are discussed in more detail. A question and guideline format is used for auditing the management systems part as well as the technical/physical conditions part. These questionnaires are translated into operational HSE programmes, which is known as 'footprinting'. This section includes examples of the footprinting process and of a graphical way of presenting its results.

The various types of monitoring activities in the systems and procedures area, as described in section 2.1, are covered in this chapter. Although the individual contributions were intended to focus on a particular category of monitoring activities within that area, they may take a wider perspective to greater or lesser extents. This demonstrates that the boundaries between the matrix sectors of Table 4.1 are not necessarily sharp: in practice, monitoring a particular sector often goes together with monitoring adjacent sectors. Moreover, there may also be overlaps with monitoring health and environment issues within integrated systems for safety, health and environment protection.

4.1   *Monitoring the implementation of local procedures*
J L Hawksley, ICI

A vital part of a safety management system is the local procedures that set down how safety requirements will be met. The implementation of the system can be described in three ways:

- what we actually do (the activity);
- what we say we do (local procedures);
- what we should do (standards, guidelines, good practice).

One aspect of good safety management is to have an adequate system and implement it effectively, so that these three descriptions are, in fact, the same. There are two types of auditing vital to achieving that (see Figure 4.1):

- firstly frequent audits that are carried out to check the activity against local procedures and answer the question 'do we do what we say we do?'. In ICI this is known as 'operational auditing' — in quality assurance terminology, these audits correspond to 'systems compliance audits'.

Figure 4.1   *Monitoring the implementation of a safety management system*

● secondly periodical audits that are carried out to check the adequacy of local procedures against standards, guidelines (both internal and external) and good practice in order to answer the question 'is what we say we do good enough?' In ICI this is known as 'specialist auditing'. It is similar to what, in quality assurance terminology, might be described as a 'systems integrity audit', but it must check that the system does cover the appropriate safety requirements as well as appropriate 'quality' requirements.

The ICI audit process also includes an overall 'management audit' to give an overall check (typically every one to three years) that the system covers all requirements and leads to improvements.

As well as the auditing of activities against the requirements of local procedures, it is helpful to track performance by measuring the attention given to safety activities on a day-by-day basis using simple positive indicators.

This section outlines some performance indicators for tracking safety activities and describes 'operational auditing' as carried out in ICI.

4.1.1   Tracking the implementation of procedures

For many of the safety activities specified in local procedures, simple indicators can be identified which give a basis for tracking performance. A few examples are the number (or %) of:

● scheduled workplace/equipment inspections completed;
● actions from inspections completed;
● process safety reviews completed;

- employees completed specified training;
- workforce involved in safety improvement teams;
- scheduled operating instructions reviews completed;
- safe/unsafe audits completed;
- reported defects followed up/corrected;
- incident investigation actions completed as planned.

Many similar simple measures can be devised for other safety-related activities. Such measures allow performance to be tracked in a variety of ways. For some parameters targets can be set to aim for, say, monthly or annually. Performance can then be charted against the targets. When setting targets it is useful to determine what would reasonably be expected to be achieved in a particular time frame and then increase that to a 'stretch target' which should be attainable but, say, 20-30% beyond what is expected.

Whilst these indicators can promote good practice and give a measure of 'what we actually do', the complementary audits are needed to confirm that what is done is done according to requirements and is of the right standard.

4.1.2   *Operational auditing*

The bulk of systems auditing (about 80%) is about ensuring conformance with local procedures. This involves checking against site instructions and obtaining maximum information about what is going on. As with other auditing activities there are important 'fringe' benefits. They give a measure of the local commitment to safety and provide a framework for targeted discussion of improvements. The discipline of auditing gets people out and about around a site, and adds to 'management visibility'.

Operational audits are regular systematic checks of an activity against local procedures and site instructions. The auditing process should be 'owned' and managed by local management with auditors mostly from local staff. All staff levels should be involved — managers, supervisors, and operations/maintenance personnel — subject to the rule that, for formal operational audit, individuals do not audit those activities that they carry out directly. Nevertheless 'self-audit' is a useful discipline that should be encouraged. The purpose of the operational audit is to determine:

- how well the SHE requirements, procedures and instructions are understood;
- the degree of conformance with the requirements;
- any steps necessary to achieve further improvement.

To achieve this purpose, auditors need sufficient knowledge and experience of the local procedures they have to check. They also need to have had some training in compliance auditing methods. The short courses, typically one to two days, offered by some quality assurance organizations can be useful training for some operational auditors.

Operational audits are carried out to an annual plan, with each element of the management system being audited at least once every two years but more often — for example, monthly — for procedures that are frequently applied or relate to more hazardous situations. There needs to be an 'audit manager' to organize and monitor the process. Local management determines the audit frequency by taking into account the nature and frequency of the activity, the consequences of nonconformance and the track record of conformance. Figure 4.2 shows an example of an annual operational audit plan illustrating how a team of six auditors covers a range of site instructions. Most audits take only a short time — say, 15-30 minutes — but longer if there are more records and details to check.

Unit: _____

Site instruction	Jan	Feb	Mar	Apr	May	Jun	Jul	Aug	Sep	Oct	Nov	Dec	Total planned	Total achieved
1		X			1			1			1		4	
2			2						2				2	
3	X	X	X	1	2	3	1	2	3	1	2	3	12	
4						2							1	
5		1											1	
6									1				1	
7	X		5		4		5		4		5		6	
8													0	
9													0	
10				4									1	
11						N/A								
12					4								1	
13		X						4					2	
14	X			6					5				3	
15		X						5					2	
16	X		X		3		6		3		6		6	
17			3							3			2	
18	X												1	
19		X			5					6			3	
20							4						1	

Auditors
1 ....
2 ....
3 ....
4 ....
5 ....
6 ....

Mark X when audit done

Figure 4.2 *Example of annual operational audit plan*

Conformance with the procedures and instructions should be judged not only by examination of critical records and the collection of other verbal and written evidence, but also by visual examination of physical conditions and methods of work. The audits should cover:

- the availability of instructions and appropriate paperwork;
- how up to date operations are in relation to safety issues;
- training and operational familiarity in relation to safety issues;
- evidence of compliance;
- practicability;
- mandatory records;
- statutory appointments;
- followup of previous audits;
- key improvements.

To facilitate operational auditing, each procedure should include an audit checklist as an appendix to the procedure. Table 4.2 gives an example.

Table 4.2   *Example of operational audit checklist*

Site instruction for pressure vessel registration and inspection — operational audit checklist

1 Locate the pressure vessel (PV) inventory. When was it last updated?	5 Check for the letter appointing the Plant Engineer as Works Engineer's nominee. Is it current?
2 Sample 3 PV files. Check the doumentation against the checklist on PVINS/1	6 Discover if there have been any new vessels on site since the last audit. Check files for PVC/2
3 Discover if any vessels have been repaired or modified since last audit. Check for completed PVC/1 and PVC/2	7 Discover if there have been any failures of PVs since the last audit. Check for a report issued by the Works Engineer for all failures
4 Locate latest overdue inspection list Check: ● Class A Vessels — Any vessel overdue should have a completed PVC/3 and the Health & Safety Executive notified of the deferment ● Class B Vessels — Any vessel should have a completed PVC/3	8 On plant tour check that PVs on site are clearly marked with their registered number

During operational audit the auditors should give immediate feedback on performance to those being audited. They should coach and advise to improve performance where appropriate. Then there should be a brief formal record of the results of each audit. Normally this should be limited to the main findings with brief comments on the degree of conformance found, a note of any particular nonconformances and, where appropriate, any prioritized suggestions for improvement. Figure 4.3 gives an example of a typical operational audit report form with a simple rating system to give a measure of performance. These reports are fed back to the management concerned who should agree and implement any specific improvement actions — the most important purpose of auditing.

Operation audited:	....
Relevant site procedures:	....
Area/plant audited:	....
Auditor(s):	....
Date of audit:	....

1	Very poor	—	immediate and extensive corrective action needed	
2	Poor	—	immediate, though selective, corrective actions needed	
3	Moderate	—	no immediate corrective action but longer-term action necessary for improvement	
4	Good	—	scope for some improvement action but not essential	
5	Excellent	—	no actions recommended	

Comments and recommendations — list Priority 1 (*immediate*) and Priority 2 (*longer term*) actions separately:

Note any site-wide problems identified requiring action:

Signed: ....   Date: ....

Figure 4.3   *Example of operational audit report form*

The 'audit manager' should collate the results from operational audits. Typical data useful for feedback to management monthly, quarterly or annually, as appropriate, are:

● numbers of audits completed against plan;
● numbers of audits reporting 'good'/'poor'/etc performance;

● average audit scores for specific procedures from successive audits to show performance trend;

● numbers of high-priority actions;

● numbers of high-priority actions not completed.

The implementation of the operational audit process should, of course, itself be subject to audit to give a measure of performance in its implementation.

4.2   Measurement of safety performance within Responsible Care
R Gowland, The Dow Chemical Company

For the last five years Dow in Europe has operated a management system audit called 'consolidated audit'. This has successfully streamlined and regularized theaudit system for safety, loss prevention, industrial hygiene, environmental, reactive chemicals and distribution.

Dow has felt that, in Europe at least, continuous improvement is important and compliance issues should be dealt with at a local level, but subject to review within an audit system. The auditors have noticed that continuous improvement is taking place, since at the second and third round of audits the follow-up has been very good. The need was felt, however, for a system of measurement.

There has been reluctance to embrace measurement systems since they can induce:

● competitive issues (plant to plant, business to business);

● score card manipulation (familiarity leads to emphasis on highly weighted items);

● problems with a moving standard (if the questions change, how do you compare with the outcome of the last audit?).

So the company is now committed to the measurement of the degree of implementation of Responsible Care.

Dow is committed to Responsible Care and has developed a system called 'operating discipline' to support it. This is a system of self-audit against a range of environmental, health and safety standards and objectives. Specialists at the location help plant personnel to complete the questionnaire and review. Validation comes later.

4.2.1   The Dow Chemical Company operating discipline
Operating discipline is the documentation and use of the collective best knowledge and experience that ensures each job can and will be performed successfully. There are 13 areas where the operating discipline management system is defined (see Table 4.3). Each of these areas is given substance and complete descriptions which allow the people operating the safety management system to see exactly where they are on a scale of 1-6. This is shown below, where management of change (section I of the Dow Chemical Company operating discipline) is discussed as an example.

Table 4.3   *13 areas of the Dow Chemical Company operating discipline*

I	Management of change
II	Process technology documentation
III	Training
IV	Safety process
V	Process risk management
VI	Operating procedures
VII	Process control
VIII	Operational reliability
IX	Dynamic process information
X	Product and service quality
XI	Environmental
XII	Industrial hygiene
XIII	Administration

4.2.2   Management of change — philosophy

Changes to facilities that produce, use, handle or store chemicals are necessary for many reasons. It is management's responsibility to assure that changes are evaluated for their effect on safety, health, loss prevention, environment and quality. The OSHA Process Safety Management Standard[1] regulates changes to chemicals, technology, equipment and facilities. The ISO 9002 Standard[2] requires that management shall plan the production processes which directly affect quality, and shall ensure these processes are carried out under controlled conditions. The Dow safety standard USA-2 'management of change' requires that changes which might affect the control or integrity of a process must be reviewed and approved.

The management of change process can be described by the following main steps:

1.   Verification: the identification of the present state of a system.
2.   Definition of change: the description of what change could affect a system.
3.   Change control: the process by which change is detected and included in the management of change process.
4.   Review: the process by which a proposed change is assessed for its impact on the process and either approved or not.
5.   Documentation: the process of changing the system's documents to reflect the change and of recording that a change has occurred.
6.   Notification and training: the process by which those affected by the change receive communications. Also includes the training on the management of change process.
7.   Auditing: the process by which compliance with the procedure for management of change is assured.

4.2.3   Management of change — glossary of terms

Six key elements are addressed in evaluating the management of change process: procedure, reviews, documents, communication, training and audits. These elements are defined as follows.

Procedure for change

The element 'Procedure' outlines the steps required to initiate, review, notify and train affected employees, approve and document proposed changes specific to the unit, department or plant. Each unit, department or plant must have a written procedure to address change to the process or facility including equipment, hardware, software, functional characteristics, physical characteristics, procedures and documentation. The procedure must define the responsibilities of the initiator and line management, required documentation, approval and reviews. The components for this key element are:

● definition of change — explains the areas covered and the types of change covered by management of change;
● levels of change — if necessary, gives attributes of the levels of change (for example: major, moderate, minor);
● authorization levels for different levels of change;
● management of change form or checklist;
● reviews or audits to assure changes comply with the procedure;
● documents which must be updated with change;
● record of the change — a document (may be the management of change form or checklist) which records the request for change and the review and approval;
● communication to affected employees.

Review

The element 'Review' describes the steps necessary to have an independent party, other than the initiator, review the concept and plans. Each proposed process change undergoes various forms of review prior to implementation. A review can be informal (one engineer discussing with another engineer) or formal. The intent of this key element is to ensure that the proposed change is fully evaluated for its impact on the process by knowledgeable individuals.
● Informal reviews — the discussion of the proposed change by the initiator with a co-worker, knowledgeable expert — for example, technology centre or plant management. The intent is to seek advice and ideas for formal reviews.
● Formal reviews — reviews by key functions (for example, safety, environment, industrial hygiene, technology) that must agree to the change prior to implementation.

Documentation

Documentation is that which will produce a permanent record of the changes which can be used for communications, training and future reference. The element 'Documentation' defines the documents that must be updated after each change is implemented. This should include but not be limited to:

• P&IDs — current set of plant P&IDs are on file and field revisions audited at least annually; (OSHA's Process Safety Management Standard[1] impacts this issue for most plants.)

• process flowsheets — the mass and energy balances for the plant;

• job procedures — the step-by-step descriptions that list the tasks to be performed, listing of known hazards and the action required to perform the task safely. Changes may cause revised procedures;

• all other operating discipline documentation — updated for all sections;

• equipment specifications — changes meet approved specifications as outlined by the plant and/or appropriate technology center.

Communication
The element 'Communication' describes the type of information and the functions that need to be informed of the change.

Training
The element 'Training' defines the individuals who must be trained prior to operating the unit with the change and the required documentation.

Audits
The element 'Audits' defines the type and frequency of audits performed to ensure that changes are properly implemented and documented.

4.2.4   Management of change — plant self-evaluation
Plant self-evaluation of management of change involves the six key elements (see Table 4.4, page 54). The scoring system is shown in Table 4.5 (see page 55). The overall unit score is the arithmetic average of the thirteen sections.

4.2.5   The validation stage
The analysis at the validation stage is done by comparing the individual elements against the management standard for each section. The management standard is mandated by the 'Governance Board' of the Dow Chemical Company. The comparison against the management standard is done during the audit programme. Follow-up is monitored by the local safety contact network.

The validation stage is covered by the Responsible Care management system audit and a description follows:

Table 4.4   *Plant self-evaluation of management of change (MOC)*

A   PROCEDURE

Question	Score						
1) There is a written MOC procedure or policy	0	1	2	3	4	5	6
2) Change is defined	0	1	2	3	4	5	6
3) Levels of change are described	0	1	2	3	4	5	6
4) Authorization levels for approval of change are defined	0	1	2	3	4	5	6
5) There is an MOC form or checklist	0	1	2	3	4	5	6
6) The MOC procedure calls for appropriate reviews for effects of the change on safety, health, loss prevention, environment, quality, regulatory compliance or impact on the customer	0	1	2	3	4	5	6
7) The MOC procedure identifies documents which must be updated	0	1	2	3	4	5	6
8) The MOC procedure prescribes how to document that a change has occurred	0	1	2	3	4	5	6
9) The MOC procedure or policy is approved by the Technical Center	0	1	2	3	4	5	6 NA

B   REVIEWS

Question	Score						
1) Appropriate reviews for effects of change on safety, health, loss prevention, environment, quality, regulatory compliance or impact on the customer are held and documented	0	1	2	3	4	5	6

C   DOCUMENTS

Question	Score						
1) Documents required to be updated by MOC are updated	0	1	2	3	4	5	6
2) MOC douments are filed	0	1	2	3	4	5	6

D   COMMUNICATION

Question	Score						
1) Affected employees are communicated with concerning changes	0	1	2	3	4	5	6

E   TRAINING

Question	Score						
1) Training for MOC is in place and documented	0	1	2	3	4	5	6
2) Training for changes is provided for affected employees and documented	0	1	2	3	4	5	6

F   AUDITS

Question	Score						
1) Annual audits are conducted to confirm that changes are in compliance with the MOC procedure	0	1	2	3	4	5	6
2) Audits for temporary changes are performed and documented	0	1	2	3	4	5	6

Total score

Average score = (Total score) / (17-NAs)

Table 4.5 *Scoring system*

0	No action planned at this time
1	Evaluating application for operating discipline
2	Developing action plan to meet operating discipline objective
3	Documentation is in progress
4	Documentation exists, is up-to-date and has been reviewed
5	Documentation has been validated and is being used
6	Documentation and use has been audited
NA	Reviewed and does not apply to the facility. Where NA is not included as a choice, the particular item is considered to be a minimum requirement and must be rated 0-6 by the facility

- It is based on Responsible Care.
- The audit format and methodology are based on the ISO 14000 series (adapting protocol used for environmental audit).
- ISO means that a management systems audit will be done.
- The audit includes deep investigation of subjects which:
  - present highest risk, or
  - cause the auditors concern, or
  - have inadequate evidence, or
  - need consultation and arbitration, or
  - require inspection or interview at workplace.
- There is a basic level of audit which applies to all manufacturing, research and development and integrated supply chain. Essentially this is the minimum standard allowed.
- There is an extra level which is elected for by each business as dictated by risk and market conditions. Examples include 'compliance audits' (see c below).
- There are three 'layers' of audits, of which a and b are basic requirements and c is optional:
  a. Local self-audit against Responsible Care management practices and operating discipline.
  b. 'Second party' audit and verification.
  c. 'Third party' audit and verification, possibly at the same time as for one of the above. This allows accreditation under EMAS[3] or some other scheme.

Second party means that the auditors are not involved in the management or control of the programme being audited. Third party means involvement of an outside public body, competent authority, consultant or community advisory panel.

Audit results are communicated primarily via the business and are locally progressed by a site implementation support person.

Follow-up

The system is administered by a global network of (second party) auditors (4-6 people equivalents). Follow-up is assured by their site level implementation support specialists (25-50% dedicated according to site scale). Auditors audit for 25% of their time. They also have regional responsibility for consulting, project review and risk analysis for the same facilities.

Audit frequency

Self-audit is a continuous process and is reviewed annually at local level. Second and third party audits are decided on the following basis:

● risk (using Dow Fire and Explosion and Chemical Exposure Indices);
● effectiveness of follow-up of previous audits;
● request by business.

The minimum and maximum periods between second party audits are three and five years respectively.

4.2.6   What is Dow's experience?

For Dow Europe the following observations for the existing audit process have been made:

● Follow-up is very good. Very few action points are missed. Discipline is generally good.
● The output from the audit process does give improvements. The second round of audits usually results in a smaller action list.
● There is output from auditing which allows changes to company standards. Improvements, upgrading and simplification have all resulted from these changes.
● Self-audit is not yet giving consistent 'scoring'. There are many more examples of people being overcritical rather than overgenerous with their self-assessment. This is usually revealed at the validation stage (second party audit).
● The audited operation finds this process a good one for interpretation of standards and 'consulting' with experts.

4.2.7   References

1.   OSHA, 1992, *Process Safety Management of Highly Hazardous Chemicals*, Title 29, Code of Federal Regulations, Part 1910.119 (Occupational Safety and Health Administration, Department of Labor, Washington, DC, USA).

2.   *ISO 9002: Quality Systems — Model for Quality Assurance in Production and Installation*, 1987 (International Organization for Standardization).

3.   *Council Regulation allowing Voluntary Participation by Companies in the Industrial Sector in a Community Eco-Management and Audit Scheme*, Regulation 1836/93, 29 June 1993 (European Community, Brussels, Belgium).

4.3   Performance measurement of safety management systems
B Fröhlich, Exxon Chemical

Exxon Chemical is continuously formalizing work and management processes to achieve further effectiveness and efficiency in the safe and environmentally sound operation of all its facilities. Work and management processes are identified, evaluated, documented, formally implemented, adhered to, maintained and improved. This leads to a formal and systematic approach in the management of safety and risk from major hazards. More specifically, in Exxon Chemical it resulted in an integrated and consistent environmental, health and safety (EHS) system called operations integrity management system (OIMS) aimed at managing all hazards and risks in the manufacturing and distribution of the chemical products and in the industrial activities involved. The OIMS addresses three areas of requirements for:

- the system (that is, what constitutes a management system?);
- the control of hazards and the management of risk;
- assessing (auditing) the overall system status and effectiveness.

The third of these is the topic of this section. It describes the process of measuring the performance of the management system — that is, assessing the overall level of implementation and effectiveness. The assessment data measured are fed back among other key environment, health and safety indicators into the risk management process of the company, in order to assure that improvements are managed, resources are allocated appropriately and the needs of the business are recognized.

4.3.1   Definition and characteristics of a management system
To evaluate and measure the level of implementation and effectiveness of the management system it is important to have a clear definition of the system. Exxon Chemical defines a 'system' within the OIMS framework of the company as follows: 'A series of steps taken to ensure that stated objectives are achieved. A typical system includes consideration of key elements: agreed objectives and documented procedures, resources responsible and accountable for implementation and execution, a measurement process to determine if desired results are being achieved, and a feedback mechanism to provide a basis for further improvement.' In this context the terms management system and system are used synonymously.

In order to meet this definition, a complete management system will have five main characteristics:

1.   Scope (boundaries) and objectives.
2.   Procedures.
3.   Responsible and accountable resources.
4.   Verification and measurement.
5.   Feedback mechanism.

These characteristics are applicable at all levels of the company, in order to achieve an integrated approach to managing hazards and risks within the line management structure.

The documentation of a system is important. All of the five characteristics need to be clearly documented and those people involved with the system should understand their role and the requirements imposed upon them. The amount of documentation should be based on the complexity of the system and be commensurate with the hazards and risks of the activity being managed.

'Verification and measurement', one of the characteristics of the management system, represents the ongoing measurement of the system, and its proper functioning is an important part in the overall evaluation of the system. Verification of the system involves confirming that the system is continuously functioning and meeting its objectives. The purpose of verification and measurement is to ensure that:

● proper functioning of the system is verified;
● processes exist to measure performance against objectives and expected results.

4.3.2   EHS hazard and risk management requirements

The purpose of an EHS management system is to control hazards and risks. In Exxon Chemical the purpose of the OIMS is to manage all hazards and the potential EHS risks in the manufacturing and distribution of the chemicals and in the industrial activities involved.

The EHS risk management expectations are laid down in a framework of Operations Integrity Management Practices (OIMPs). These practices contain key risk management requirements for:

● management leadership, commitment and accountability;
● risk analysis, assessment and management;
● facilities design and construction;
● process and facilities information and documentation;
● personnel safety;
● health;
● personnel;
● training;
● operations and maintenance procedures;
● work permits;
● inspection and maintenance;
● reliability and control of defeat of critical systems and devices
● pollution prevention;
● regulatory compliance;
● product stewardship;

- management of change;
- third party services;
- incident reporting, analysis and follow-up;
- emergency preparedness;
- community awareness;
- operations integrity assessment and improvement.

The management system must address the requirements and objectives proportionate to the risks of the operation.

4.3.3   Assessment of the management system

In Exxon Chemical there are two steps involved in the overall assessment of the safety (or integrated EHS) management system:

- the self-assessment of the system;
- the external assessment of the system.

The self-assessment is an in-depth audit of the system and the activities required by the system, and is completed annually. The formal assessment procedure requires credibility of the team involved in the auditing of the activity being assessed. In smaller operations this is ensured by inviting experts from outside the unit. The leader of the self-assessment is a member of the site or unit management team. This assessment is aimed at measuring key indicators of the system:

- The system is under development.
- It is documented, approved, resourced, and being implemented.
- It is functioning and being verified; performance indicators are established and measured at appropriate levels; key system procedures are documented.
- It is sustainable and supported by an ongoing improvement process.
- It effectively meets specific risk management objectives defined by OIMS.

Thus, the level of implementation and the effectiveness of the safety management system are evaluated in detail, using a standard format. A numerical level is applied indicating the degree of meeting system and risk management expectations of the company. Written comments are made to indicate actions needed for a complete and effective system.

The external assessment is carried out once every three years by an experienced multi-disciplinary team led by a senior manager. The entire team is external to the operations — that is, coming from other units or sites — and is mostly an international team. The key qualities of the team members are:

- experts from central functions (EHS/other engineering specialties);
- members of operations management from similar or larger operations;
- experienced assessors for continuity.

The purpose of the external assessment is to verify and validate the self-assessments. That is, it confirms or corrects the level of performance measured by both the numerical 'score' and the comments made for improvements.

4.3.4   Evaluation of the management system

In evaluating the system during the self-assessment, the first step is to examine system documentation, in order to determine whether the five basic characteristics of a management system are incorporated. Identifying the five characteristics allows an assessor to conclude that the system 'design' is adequate for the EHS hazards of the operation.

Adequate design alone does not control the risks in an operation. The actual application of the system — for example, the way in which risk management requirements are addressed — must be tested. There are two dimensions that are necessary for a good management system and should be reviewed to evaluate a system: status and effectiveness. Evaluation of these two important dimensions, which are discussed below, is the key to judging the suitability of a system in use in an operation.

4.3.5   Evaluation of the management system — status

The status of a system is a measure of the stage of development and implementation of a system in the operation. Figure 4.4 shows four stages of system status.

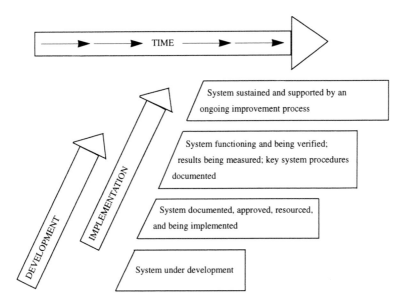

Figure 4.4   *Stages of system status*

In the early stages, shown in the bottom box of Figure 4.4, a formal system is being developed to improve the overall EHS management system of a related group of activities according to the system and risk management expectations.

In the next stage, the system is introduced into the operation. At this stage, the system (including system process steps) is documented, approved and resourced, and implementation is underway.

In the third stage, the system is functioning in the organization. Adjustments to the documented system process steps, if required, have been made for completeness and to ensure that the system functions as intended. Also, ongoing verification measures indicate that it is working as intended. Procedures for key system tasks are documented. Key results and outputs are being measured.

The fourth stage is a fully mature system. Resources are in place to sustain the system and the system has undergone at least one feedback/review/improvement cycle. It is important to note that at this stage a continuous improvement process is in place to identify and execute improvements to the system. The continuous improvement loop is required to sustain the system over time. It ensures that the system continues to operate effectively with organization and facility changes.

4.3.6   Evaluation of the management system — effectiveness

The other dimension used to evaluate an applied system is effectiveness. The system documentation describes the scope of activities/operations and risks to be managed by the system, and the hazard and risk management results expected. In this context, the activities/operations, risk management requirements and results are called 'system objectives'. The effectiveness dimension is a measure of the extent to which the system objectives are satisfied. Figure 4.5 shows four categories of system effectiveness.

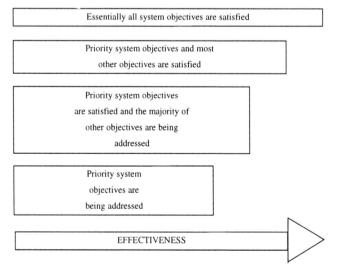

Figure 4.5   *Categories of system effectiveness*

Categories of system effectiveness

A system will normally satisfy priority objectives and other (secondary) system objectives. If priority objectives are being addressed but are not satisfied, the lowest category of system effectiveness is indicated (bottom box of Figure 4.5). If the priority objectives are satisfied and a majority of the other objectives are only being addressed, the second category of effectiveness is indicated. When the priority objectives and most other objectives are satisfied, the third category is indicated. The fourth and highest category is indicated when essentially all system objectives are satisfied.

Quality

The overall quality of the system must also be evaluated as part of the effectiveness dimension. Table 4.6 presents quality considerations for systems. Quality is defined here as the 'suitability of a system for its purpose'. A judgment of a system's suitability is required.

Table 4.6 *Quality considerations for systems*

TIMELINESS

    System outputs are produced and acted upon in a timely manner

    Deviations / exceptions to expected results are handled in a timely fashion

    Verification is conducted at regular, appropriate intervals

    Backlog of system outputs is at an appropriate level

ALIGNMENT

    The system responsibilities and objectives have been communicated, understood and accepted by those involved with the system

    The system is being used consistently

ADEQUACY OF RESOURCING

    Appropriate resources have been allocated to implement and sustain the system

    Sufficient training has been conducted for those involved with the system

---

ADEQUACY OF OUTPUTS

    Appropriate emphasis is placed on higher potential risk activities

    Information generated by the system is pertinent, accurate and thorough

    System outputs are presented in a meaningful fashion

4.3.7   *Overall evaluation of the management system*

It is helpful to draw the analogy of an umbrella to illustrate the status and effectiveness dimensions. The status dimension would be concerned with having (or not having) the umbrella in your hand when it starts raining. The effectiveness dimension would be concerned with how far the umbrella is opened and how well it keeps you dry. An umbrella that is only partially opened or leaks would not keep you dry, and its effectiveness would therefore be low.

Assessing the effectiveness of a system against what is appropriate in an operation is a judgmental exercise and requires an experienced assessor and assessment team. The overall evaluation of a system includes system design, implementation status and system effectiveness. The process for this evaluation is shown in Table 4.7.

Table 4.7   *Overall system evaluation process*

STEP	ACTION
1	Evaluate system design and implementation status (against five characteristics)
2	Evaluate system effectiveness
3	Identify areas for improvement or 'gaps'
4	Determine overall system level

In reviewing system design (step 1), the system is evaluated for completeness against the five basic system characteristics. The implementation status of the system is evaluated for addressing the priority risk management objectives (OIMP).

In step 2 the system effectiveness is evaluated based on the actual application of the system versus the objectives of the OIMPs.

Step 3 provides the feedback to the assessed unit for improvement of their systems. The improvement areas for systems already in place are identified, and also the gaps between the actual system effectiveness and the system requirements. Depending on the assessment protocol, recommendations for the identified improvement areas and gaps may be provided (system design and/or risk management requirements).

Step 4 involves determining the overall system evaluation. When combining the status and effectiveness dimensions, the lower level of either dimension determines the rating (score) level for that system. Table 4.8 describes this relationship in more detail.

Table 4.8   *Dimensions to consider in overall system evaluation*

SYSTEM STATUS	SYSTEM EFFECTIVENESS
System sustained and supported by an ongoing improvement process	Essentially all system objectives are satisfied
System functioning and being verified; results being measured; key system procedures documented	Priority objectives and most other objectives are satisfied
System documented, approved, resourced, and being implemented	Priority objectives are satisfied and the majority of other objectives are being addressed
Under development	Priority objectives are being addressed

4.3.8   Conclusion

Exxon Chemical has an integrated environmental, health and safety management system (OIMS), which contains expectations (requirements) related to the system, the control of hazards and risks, and the assessment of the overall system status and effectiveness. The assessment or auditing process is formalized and has a three-component measurement approach:

● measuring continuous improvement (ongoing verification, key performance indicators established);
● measuring performance of the overall system (yearly self-assessment);
● verifying and validating the above (three-yearly external assessment).

A numerical classification level (score) is used in the overall assessment of the system, rating the key indicators of the management system and the level of meeting hazard and risk management expectations. Assessment results are communicated to company management as input for overall improvement of the company's risk management process.

4.4   Measurement through auditing and assurance
R G Read and R G Yeldham, BP

The BP group is an integrated hydrocarbon company operating in more than 70 countries worldwide, with its operations divided into three core businesses, namely upstream exploration and production, downstream refining and marketing, and chemicals. It has an annual turnover of £36 billion, and an international workforce consisting of 56,000 employees and 40,000 contractors.

4.4.1   *Integrated health, safety and environmental management*

Health, safety and environmental (HSE) matters have long featured as issues of importance in BP. These topics have always been managed in a related manner, but in the last few years they have become even more closely integrated.

The principle of the approach is to manage the operations well, and avoid losses of all kinds. Doing that, and preventing, say, the release of hydrocarbon to the atmosphere will clearly benefit the safety (no fire or explosion) and health (no long-term effects) of the workers, and the environment (no spill). BP has also demonstrated quite categorically that the absence of spills, damage and business interruption makes good economic sense, and recognizes that the opposite side of the coin — failure to meet the standards in HSE demanded by regulatory authorities and the community generally — may cause the company's 'licence to operate' to be withdrawn.

4.4.2   *HSE management system*

The whole approach to HSE is driven from the very top of the company. The HSE management system is shown in Figure 4.6, page 66. It starts with the HSE policy, issued by the main board of the company, and signed by the group chief executive. As it happens, the policy has just been re-issued in January 1996, following a major review commissioned by the incoming chief executive. This was driven by:

● increasing importance of getting the management of these issues right;
● increased emphasis on the responsibility for HSE management being in the line.

The previous policy had served BP well, but there was concern over its clarity and relevance to the new structure of the company. An employee telephone survey, a line manager workshop, and ultimately the opinions of over 1000employees and contractors round the Group were sought. The result was that a much simpler and clearer statement of the company's commitment to health, safety and environmental performance was generated (Figure 4.7).

Expectations

Below the policy a short list of high-level expectations was developed, which describe the deliverables which the Business Units round the world will be measured against. From them are derived the processes, procedures, operating standards and other programmes by which each unit delivers its business performance.

BP's approach is to integrate HSE management into all the company's activities so that everyone appreciates that it is their responsibility. HSE requirements are agreed within teams and built into the appropriate site, group and personal objectives. HSE programmes are built into the normal business planning processes, and into performance contracts. Line managers may choose to be assisted by professional HSE advisers, but HSE is not seen as a function that should be managed separately from business operations.

Each Business Unit manager has in place systems for continual review and periodic audit from which to make an assessment and measurement of programme implementation. This assures the manager that things are under control; it can also be used in reviews with the manager's peers, and to provide upwards assurance within the company.

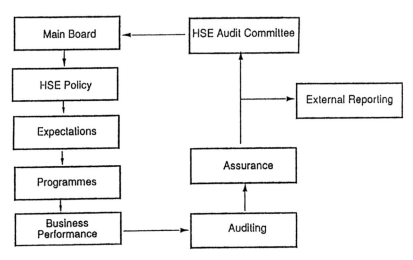

Figure 4.6   *HSE management in BP*

Assurance
BP carries out a formal annual assurance process in which each Business Unit manager provides a written report describing:

● the management of the key HSE risks under the manager's control;
● compliance with legal requirements;
● progress with implementation of the policy;
● progress against performance targets and objectives;
● the programme in place for the next year to continue the process.

These reports are transmitted upwards through the management line. The final product is a report from the group chief executive to the HSE audit committee, which is a formally constituted part of the board, consisting solely of non-executive directors of the company. They meet three or four times a year, and review all aspects of BP's HSE performance, including this annual assurance process. They also review more detailed reports from different parts of the company on a sequential basis and, from time to time, commission group-wide audits on HSE themes of particular importance.

BP's commitment to health, safety and environmental performance

Everybody who works for BP, anywhere, is responsible for getting HSE right. Good HSE performance is critical to the success of our business.

Our goals are simply stated – no accidents, no harm to people, and no damage to the environment.

We will continue to drive down the environmental and health impact of our operations by reducing waste, emissions and discharges, and using energy efficiently. We produce quality products that can be used safely by our customers.

Wherever we have control or influence we will:

- consult, listen and respond openly to our customers, neighbours, and public interest groups

- work with others – our partners, suppliers, competitors and regulators – to raise the standards of our industry

- openly report our performance, good and bad

- recognise those who contribute to improved HSE performance

Our business plans include measurable HSE targets. We are all committed to meeting them.

John Browne
Group Chief Executive

HSE Policy
January 1996

Figure 4.7   *BP's commitment to health, safety and environmental performance*

Closing the loop

The acceptance of the assurance report closes the loop, and the process continues. The reissuing of the policy is one indication that BP has a live system, since it was — amongst other things — the action arising from the 1994 report, proposing improved clarity on the company's commitment to eliminating accidents, that led to the review just described.

4.4.3   Workplace HSE auditing
The main focus of this section is the auditing process used at the workplace to review and measure HSE activity. The example which is used is the process adopted by Business Units in BP Oil, the downstream part of BP, which produces, distributes and markets oil fuels and lubricants to industry and the public world-wide.

4.4.4   The BP Oil approach
BP Oil Business Units are responsible for continually reviewing and measuring their own performance against the expectations and their own site-level requirements. There is a strong focus on HSE auditing, which has always been seen as fundamental to that process. A large number of such audits has always been carried out each year, and agreed audit recommendations are followed up until closed out by accountable management. Audit coverage of the whole area of HSE risk has generally been seen as acceptable, but more recently BP Oil has seen reason to take a much more structured and systematic approach to the whole topic.

History
In the early 1990s BP Oil, like many other companies, suffered from a surfeit of audit types. These included: legal compliance audits, contractor audits, emergency response reviews, International Safety Rating System audits, HSE strategy reviews, major hazard reviews, joint HSEQ audits, product stewardship audits, general safety and health reviews, environmental reviews, occupational health reviews, fire audits and technical safety audits.

There were overlaps and gaps in audit coverage, a multiplicity of uncoordinated audit recommendations in different HSE languages and styles, an imbalance between the number of audits of different types, poor use of auditor resources and inconsistent or poor quality audits. Line managers felt overwhelmed by the number of audits and saw some as unconstructive exercises thrust upon them from on high, having little bearing on the management of the real risks at their site. At the same time there were increasing demands for companies to satisfy various external audiences about their management of HSE, and the companies had their own need to provide evidence that their systems could pass a number of external 'tests'. And all this within an increasingly difficult commercial environment.

4.4.5   The way forward — REALM
A global, cross-business team looked at the problems and developed a solution which not only removed many of these difficulties but also helped to integrate HSE further into line management processes and thinking.

The proposed framework was a 'one stop' system which delivered: a means of auditing, a tool to help managers set HSE plans and targets, and a mechanism to facilitate sharing and comparing between Business Units. The framework is known as REALM (resource efficient auditing for line management).

The framework has two parts (see Figure 4.8):

1. A management systems component (REALM1) essentially common across all elements of HSE.
2. A technical/physical conditions framework (REALM2), subdivided according to the HSE disciplines. The latter addresses competency issues and the technical adequacy of both procedures and equipment.

REALM1 and REALM2 are presented in the same style, and use the same question and guideline format. Both aim to be comprehensive, encompassing not only best practice, but also a range of lesser although still 'fit for purpose' approaches.

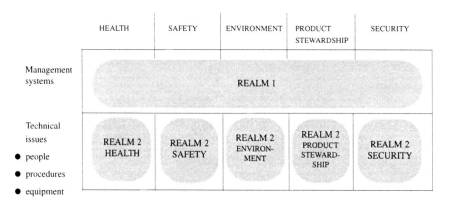

Figure 4.8 *Business unit HSE management*

The most important single feature of the system, however, is not its detail but the way it is applied. The concept is that Business Unit managers should be primarily responsible for determining what is right for their business and for selecting/discarding activities from the master document set accordingly. This process helps them develop their own plans and provides a template for auditing. This is known as 'footprinting' and is described in more detail below.

Development of REALM

The platform for REALM1 had to be either a purpose-built system or an extension of an existing system, as no current system could be found which covered the necessary range and had the right structure. BP already subscribed to DNV's International Safety Rating System, and after a wide-ranging review it was decided to use this as a core for REALM1. However, the set of elements was expanded to encompass best practice, each was applied across the whole range of HSE issues, and greatly enhanced guidance was provided based on the company's own experience. In addition the auditing and verification procedures were tightened up considerably.

A set of comprehensive REALM2 audit modules for refineries has also been developed and is nearing completion. This document set is based entirely on general oil industry practice or BP Oil's own experience. Parts have been in place for some time and have proved effective.

4.4.6   Footprinting

The main purpose of REALM is to help the business meet its HSE and other goals by matching resources to risk and by providing evidence that regulatory and industry practices are being met.

This is achieved by applying the risk management process known as 'footprinting' to the master REALM documentation. Footprinting is the process of translating REALM questionnaires into operational HSE programmes. It is fundamentally management's job to understand and define the programme and to articulate it in as concise a way as possible, to gain the widest involvement and understanding.

A team from the Business Unit assesses the relevance of each REALM activity to the risk profile of their own operation. Footprinting involves discarding the irrelevant or non-cost-effective methods, and ranking the remainder, incorporating them into either current activities or forward plans for the years ahead.

As well as risk, the team takes into account the goals of the business and any 'givens' such as legal compliance and codes of practice to which the organization subscribes. Increasing importance is being put on compliance with external industry or international (ISO) codes. Examples include the European Eco-Management and Audit Scheme and the Responsible Care principles. Generally such codes contain little that is not already accepted practice and therefore they can easily be mapped on to REALM and be built into the appropriate footprint. For instance, by taking the REALM2 environmental module together with an appropriately footprinted version of REALM1, BP Oil has successfully piloted in Europe a very cost-effective, systematic approach to eco-auditing at one of its refineries.

Ideally the footprinting process requires the involvement of a management team knowledgeable about the hazards present, the safety principles to be followed and the techniques available, as well as having the necessary support from specialists. To this team should be added those with hands-on knowledge of the work involved and representatives of all those who will be concerned in programme implementation. In reality such a blend is difficult to achieve, but these principles are applied as far as is practicable.

A simplified example of the footprinting process is shown in tabular form in Table 4.9. It is drawn from elements 5 and 9, accident/incident investigation and analysis respectively. Within these two elements there are about 90 possible requirements in the form of questions and supporting guidance. Each one has been considered by the footprinting team for the hypothetical Business Unit, and Table 4.9 shows the outcome of their deliberations for just five of those questions.

Table 4.9 *Example of footprinting process*

REALM	ACTION			
QUESTION/STATEMENT	Now	Next year	Two years	Never
1. Is there a system for reporting and investigating injuries, occupational illnesses, non-conformances, property damage, environmental damage/complaints, security breaches and other incidents including near-misses?	✓	-	-	-
2. Are team leaders trained in investigation methods?	✓	-	-	-
3. Are costs of accidental loss collated and analysed?	-	✓	-	-
4. Are alternative records, such as first aid log books, insurance claims etc reviewed to assess the effectiveness of the incident reporting system?	-	-	✓	-
5. Are team leaders given training in advanced problem-solving techniques?	-	-	-	✓

The team has decided that two requirements should be in place now: one for a system for reporting and investigating injuries, damage etc, and one for team leaders to be trained in investigation methods. It has also decided that a cost collection system should be in place next year, that a system to review alternative records to cross-check effectiveness should follow a year later, but that advanced problem-solving training for team leaders is inappropriate for the risks in their business, and will never be required. The footprint will show these requirements, and the management audits will measure progress in achieving them.

The benefits of this process are many but include:

● an open, transparent process involving 'bottom-up' ownership, and hence greater chance of implementation, of a set of well-tried systems;
● accreditation against external management systems requirements with minimum internal disruption;
● systems tailored to need.

A full footprint, containing all necessary qualitative detail, is in the form of an annotated subset of the 'master' REALM1 questionnaire. The use of a common REALM1 framework for all footprints worldwide therefore facilitates 'sharing and comparing' across the BP Oil businesses.

4.4.7   Performance measurement

Although each footprint is compiled question-by-question, it can be expressed graphically — in the form of a bar chart — by taking account of the scores attached to each question (Figure 4.9). This can show the targeted performance of the particular business stream over a number of years, and across a range of activities. Audit results can be expressed in the same form. Footprints are the key to measurement, and are suitable for carrying out peer reviews, or for inclusion or reference within business plans and performance contracts.

4.4.8   Summary

Proactive measurement of safety performance (or health, safety and environmental performance) has to be an integral part of the management system. Every company will be different, but BP has found that as the organization evolves, so does the management system, and clarity of roles for HSE as for every function is essential. Strong management ownership of the system enables the policy to be implemented in a consistent fashion, and robust methods of auditing and assurance to develop. The development continues; it will probably never end, and the ability to measure performance on a regular and consistent basis provides the essential feedback to keep the system alive.

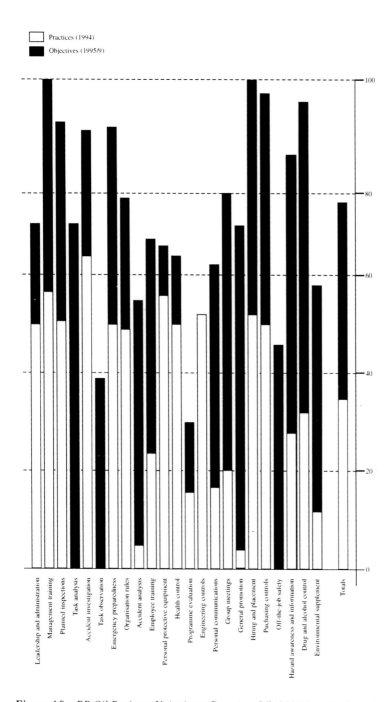

Figure 4.9 *BP Oil Business Unit (Asset Group) — REALM1 footprint bar chart*

5 Measuring people: practical examples

The third area of safety management inputs is concerned with people who are competent, through knowledge, skills and attitudes, to operate the plant and equipment and to implement the systems and procedures (see Table 5.1). Monitoring activities in this area include behavioural observation and feedback, attitude surveys, and overall management audits.

Table 5.1 *Safety performance measurement — monitoring the inputs*

	Regular inspections and audits by local staff	Periodic in-depth inspections, assessments, audits of specific aspects by specialists	Overview assessments and audits by independent experienced assessors
Plant and equipment			
Systems and procedures			
PEOPLE			

This chapter presents five examples of monitoring activities in the people area. To begin with, the behavioural approach to safe working at BASF is described, which deals with observing behaviours on the shop floor. After introducing the approach and indicating the reasons for choosing it, section 5.1 deals with the four steps of implementation by presenting an actual application. Results before, during and after a major plant overhaul are given. The section concludes with an overview of barriers which may affect implementation of the behavioural approach to safe working.

DNV's contribution (section 5.2) addresses measurement of behaviour of people in leadership as well as operating positions. Following the presentation of a conceptual view on behaviour measurement, three categories are described: the evaluation of 'leadership behaviour', measuring general behaviour of people in operating positions, and task observations of people in operating positions. The first of these categories is addressed through interviews, the other two through observations of behaviour. The section includes examples of questions for leadership performance.

Next, AEA Technology describes a tool for assessing safety culture (section 5.3). The tool is derived from a safety culture framework, in which 129 safety culture parameters are hierarchically organized into three main areas: the management and organizational level, the enabling activities and the individual factors. Assessing the first two of these areas is done through interviews, whereas the individual factors are assessed by questionnaire. Examples of questions are given. The case studies in which the tool was applied were concerned with the nuclear sector, the rail sector and the oil and gas sector.

The contribution from Shell in section 5.4 discusses the Tripod incident analysis methodology which distinguishes three 'feet' in incident causation: the incident itself, active failures and latent failures. The Tripod methodology categorises latent failures into eleven general failure types (GFTs), such as 'training', 'hardware' and 'communication'. It can be used reactively in incident analysis and proactively as a diagnostic tool for accident prevention. In both cases, results are presented in a so-called failure state profile: a histogram of GFT frequencies.

The final section of this chapter is concerned with the DuPont approach to safety performance measurements. These recognize the distinctions between plant and equipment, systems and procedures, and people, but the 'people' aspect is viewed as an integral part of all aspects of the business. Thus, the DuPont system of process safety management (PSM) is briefly explained. Four steps to effective PSM are presented, followed by a description of safety performance measurements which include both lagging and leading indicators. Leading indicators which are discussed are the unsafe acts index, the PSM scoring system and the process incident scoring system. The section concludes with some safety performance results.

The various types of monitoring activities in the people area, as described in section 2.1, are covered in this chapter. Although each individual section was intended to focus on a particular category of monitoring activities within that area, they may take a wider perspective to greater or lesser extents. This demonstrates that the boundaries between the matrix sectors of Table 5.1 are not necessarily sharp: in practice, monitoring a particular sector often goes together with monitoring adjacent sectors. Moreover, there may also be overlaps with health and environment monitoring within integrated systems for safety, health and environment protection.

5.1   *The behavioural approach to safe working*
G C W Walker, BASF

BASF plc at Seal Sands is part of the world-wide BASF group of companies. It is a large chemical complex on the north bank of the River Tees, near Middlesbrough, in the north-east of England. The site occupies just over 100 hectares, and manufactures acrylonitrile, adiponitrile and hexamethylenediamine, which are used in the production of acrylic and nylon 66 fibres and a wide range of engineering plastics. It is a 'top-tier major hazard' (CIMAH) site.

The original plant was built by Monsanto in 1969, and the site was significantly extended in 1979. The original plant was closed in 1981. BASF purchased the site at the end of 1985.

In 1996, the site employs about 400 BASF and 300 contractor personnel, many of whom have been employed at the site continuously through both ownerships. All the routine maintenance and many other services are carried out by the contractors who are included in the site's safety statistics.

BASF plc subscribes to the UK Chemical Industries Association, and is active in many international groups concerned with safety and environmental protection.

5.1.1   Introduction to the behavioural approach

The way people behave has a significant influence on the potential for accidents. The actual frequency and severity of such accidents has a large dependence on 'luck' - milliseconds or millimetres can make the difference between a 'near-miss' and a fatality.

Some accidents may be caused by unsatisfactory conditions at the time, but most occur when everything is apparently up to standard. Often accidents are not linked to the specific hazards of an industry. Most of the accidents can be avoided by an increase in personal care, but simply telling people to take more care does not work. It is also becoming more difficult to make progress through direct legislation.

The behavioural safety approach identifies, emphasizes, measures and promotes 'safe' behaviours rather than the punishment of 'at risk' behaviours. It works best when the people at the workplace believe that 'safe' behaviour is the acceptable norm, and is not unreasonable or cowardly, nor to be ridiculed, and where 'at risk' behaviour is corrected but not punished.

An HSE-commissioned study ('The development of a model to incorportae management and organizational influences in quantified risk assessment', HSE Contract Research report No 38/1992, published by the HMSO in London) traced the development of improving safety performance from a traditional approach of believing that workers are careless, need to be told what to do and should be punished for not doing it, through a procedural/engineering period in which protective devices and safety management systems were developed and implemented, to a behavioural approach which treats workers as mature human beings who work best when consulted and involved in how tasks should be done.

If people are correctly taught and 'encouraged' to behave safely, this behaviour will become a habit which will be followed, no matter the circumstances. The challenge is to find data which allows proactive action; observations and measurements of behaviour give such data.

It is a well-established safety approach to relate frequencies and severity of injuries, and it is simple to extend the traditional 'triangle' to include near misses and behaviours (Figure 5.1). At risk behaviour is an early warning system for accidents.

5.1.2   Why did BASF choose the behavioural approach at Seal Sands?

There were several reasons why the site chose the behavioural approach:

● The safety performance at the site had 'plateaued' (Figure 5.2).
● The approach had been described as appropriate in the HSE report mentioned above.
● The system had given positive results in another location with a similar culture.
● The system was identified as a 'continuous improvement' safety process, and would not conflict with other initiatives.
● It was an initiative which management could support but which would be led from the

workplace by those most likely to experience injuries, and one which could empower employees to contribute significantly to improvements in their own safety.

● As a 'no fault' system, it could be a mechanism for more open exchange, breaking down known barriers between supervision, operators and contractors.

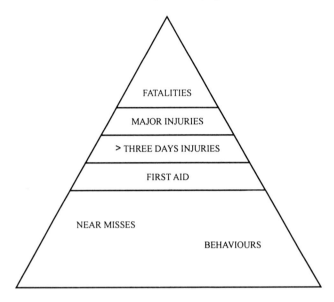

Figure 5.1 *Extension of traditional accident 'triangle' with near misses and behaviours*

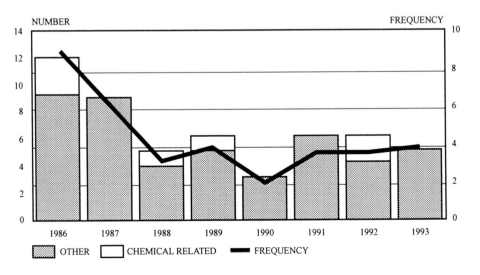

Figure 5.2 *Safety performance (reportable injuries per 10^6 man-hours) of BASF plc at Seal Sands*

5.1.3   What are behaviours and how are they influenced?

Behaviours are observable acts. They can be seen! A multitude of factors influence people's behaviours. Examples include:

- they may not know the correct way to behave;
- they may not have the correct equipment/tools to perform the task;
- they may be subject to other pressures — for example, from supervisors or peer groups;
- they may not understand the consequence of their action;
- they may not have been trained.

Things which set the scene around a behaviour (precursors, triggers, or antecedents) and the consequence of the behaviour (consequences) both influence the behaviour. The consequences are the more significant. Consequences which are immediate, certain and positive influence behaviour more significantly than those which may come later, are uncertain and negative.

5.1.4   Experience at BASF at Seal Sands

BASF decided to adopt the process developed by Behavioural Science Technology Inc (BST). A steering committee of volunteers from among the BASF operators, various contractors and some services groups began training in July 1994. In keeping with the principles of leading from the shop floor, the chairman elected by the committee was a process operator. Supervision is represented on the committee by a BASF shift supervisor and a contractor supervisor, and management support is through the safety manager as a facilitator.

The steering committee was given the task of guiding the process through the four steps of implementation:

1.   Identify and define critical behaviour.
2.   Train observers to recognize and measure the occurrence of the behaviours in the workplace.
3.   Establish a system of ongoing observations and feedback.
4.   Use the gathered data to identify corrective actions and plans for continuous improvement.

Step 1: Identifying and defining behaviours

Under the direction of a BST consultant, the steering committee members analyzed injury reports to identify the cluster of safety-related behaviours that had contributed to past accidents and sorted these into generic types. Included on this list were behaviours such as bending correctly, using appropriate protective clothing, and using the proper tools for a job.

Once the critical behaviours were identified, the Steering Committee wrote a list of operational definitions for them and produced a data sheet for measuring their frequency on the shop floor. Table 5.2 shows a typical definition.

Step 2: Observer training

The next step was to train observers not only to use the data sheet to recognize and record 'safe' and 'at risk' behaviours in the workplace, but also to let the workers know how their work practices measure against the operational definition. This technique is not as simple as it may seem — the observations must be seen to be no threat to the person(s) being observed, and the feedback must be presented with a great amount of tact.

The steering committee trained the observers to accomplish these observation goals by:

● announcing their presence to workers that they observe;
● showing the workers that the data sheet has no place for names, and reassuring them that their names will never be attached to the observation;
● sharing the results of the observation with those observed before entering the information into the observation database;
● giving positive, encouraging feedback to those observed;
● reporting the 'safe' behaviours observed first;
● asking questions as to the reasons why the 'at risk' behaviour occurred instead of criticizing the worker.

Table 5.2 *Example of critical behaviour and its operational definition*

EYES ON HANDS/WORK
Are eyes on work/hands at all times?
eg: does the person stop work when distracted?
Is care taken when using striking or cutting implements?
eg: hammer, drill, grinder etc
Is care taken when handling chemicals?
eg: sampling

After an initial period of uncertainty, the workers were convinced by this straightforward approach and are now very supportive of the process.

Step 3: Continuous peer-to-peer observation and feedback

By March 1995, an initial number of shopfloor personnel had been trained as observers, but observations were not being conducted at the desired frequency. The underlying frequency should be such that each person is observed at least once a month. But progress was slowly being made.

In September 1995, the process was given a boost during the major overhaul of the nylon intermediates plant. The overhaul involved the co-ordination of 1429 jobs and 7054 activities, to be carried out by approximately 450 people, including 300 temporary personnel employed by more than 50 companies.

In keeping with the large maintenance effort, observations were scheduled as part of the work load. A total of 426 observations were carried out during the 23-day overhaul.

Table 5.3 shows part of the analysis. Other reports allowed the quality and quantity of observations to be monitored.

Members of the steering committee were involved in the induction of the temporary contractors, and were able to generate high levels of interest and participation from them. There were reservations at first, but these were short-lived when the temporary workers realized that the process was the result of a genuine interest in their safety and that it posed no threat to them.

Table 5.3 *Personal protective equipment: analysis of observations during major overhaul*

		SAFES	AT RISKS	% SAFE	SHEETS
3.1	Eyes/face	386	36	91	417
3.2	Hands	344	94	78	416
3.3	Feet	404	18	95	421
3.4	Head	397	16	96	410
3.5	Hearing	44	56	44	84
3.6	Breathing	27	25	51	45
3.7	Protective clothing	305	20	93	324
3.8	Safety harness	3	7	30	7
		1910	272	87	

Step 4: Use gathered data to identify corrective actions and plans for continuous improvement

The final step of the process is to analyze the observation data for barriers to continuous safety improvement, remove the barriers and install positive consequences for safe behaviour. Actions are assigned to those best suited to carry them out. The result is a continuous safety mechanism that is proactive and data driven.

The analysis of observation data gathered during the shutdown highlighted areas for focus. Examples were the availability of safety helmets that are compatible with ear protection and visors, difficulties in access, working with pipes, kneeling on uneven surfaces and many more. As a result of these findings, Seal Sands has implemented several changes in operating methods, ways to achieve access, and types of equipment available. In particular, portable vices and pump lifting frames have been brought into use.

During the shutdown, 22 minor injuries were reported — more than a 30% reduction on the nearest comparable overhaul statistics. All of the injuries were from low-scoring areas on the critical behaviour data sheet. Being in 'line of fire' was observed at 59% safe (15 injuries), 'eyes' on path' at 69% (5 injuries) and 'restrictive access' at 55% (2 injuries) (see Figure 5.3). During the shutdown, the rolling percent safe rose over the 23-day period from 75% to 82% (Figure 5.4). These facts demonstrate both the value of the critical behaviour list as a continued area for focus and the benefit of conducting frequent observations.

Subsequent to the overhaul, the number of regular observations dropped back to pre-shutdown levels, but they are steadily rising (Figure 5.5; see page 82). With the increase in regular observations, the overall site percent safe has risen to 80%, and the site has achieved the longest run without a reportable injury since start-up in 1969 (Figure 5.6; see page 82).

To continue with improvements, more information needs to be gathered by observations. Targets have been set, and more site personnel are being involved. Contractors have been as supportive as BASF direct employees, with the process of behavioural accident prevention giving everyone the opportunity to participate and improve the safety performance.

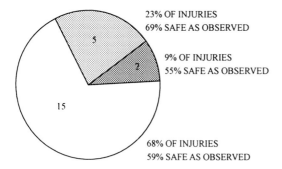

23% OF INJURIES
69% SAFE AS OBSERVED

9% OF INJURIES
55% SAFE AS OBSERVED

68% OF INJURIES
59% SAFE AS OBSERVED

Figure 5.3 *Behavioural causes of accidents during 23-day shutdown*

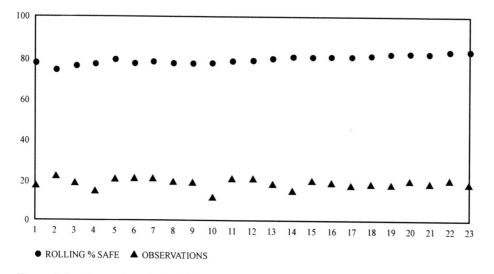

● ROLLING % SAFE   ▲ OBSERVATIONS

Figure 5.4 *Observations during 23-day shutdown*

5.1.5 Barriers

There are a number of barriers which may affect implementation of the behavioural approach to safe working.

● The safety or company culture and commitment: There is a commitment to time and effort which amounts to 1.5%-2.5% of time. The approach is long-term — say, 2-5 years.

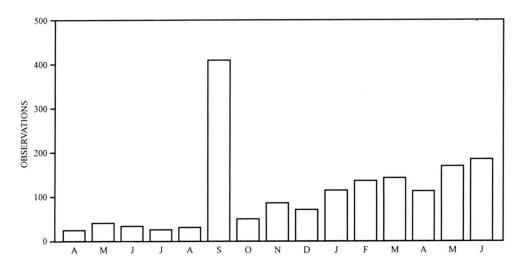

Figure 5.5   *Monthly observations*

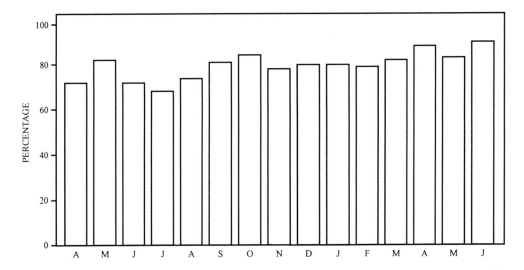

Figure 5.6   *Monthly percentage safe*

● Plant and equipment may be in itself unsafe: This indicates poor commitment to safety.

● Individuals at all levels feel threatened: The behavioural approach is new in many aspects. Skilled personnel think that people not skilled at the job being observed cannot make proper observations. This remains until it is understood that it is not technical content that is being observed but only behaviour. Supervisors, union officials, safety committee members and others may feel their authority is being undermined.

- Management systems may need to be changed: Management must be shown to be supporting this approach and not enforcing it. Procedures may need to be changed.
- Payment and reward schemes may conflict: It is possible to run schemes which reward safe behaviour. But it should be possible to achieve sufficient buy-in to eliminate the need for reward schemes.
- Disagreement on safe practices, or the need for improvement: The use of the list or inventory of behaviours critical to safety means that the behaviours are defined, and agreed before any observations are made.
- Personal choice may be affected: It may well be! If the approach is to succeed, each individual must be convinced. This can take time and effort but is well worthwhile to get individuals to buy in.

5.2   *Measurement of behaviour*
W N Top, DNV

The measurement of behaviour is, in principle, directed at two groups: people in leadership positions (managers, supervisors and relevant staff) and people in operating positions.

People in leadership positions include all persons who, through their acts and doings, influence the behaviours of others. Through their example, what they do or do not do, their behaviour and decision-making, the organization makes clear what it finds important. The measurement of 'managerial' behaviour is crucial here, since the behaviour of these people greatly influences other people's behaviour (and the consequences thereof).

The measurement of behaviour of people in operating positions can be divided into two categories:

- (general) behaviour observation, which is the observation of the way people behave in a more general sense while doing their work;
- task observations, which are directed at the performance of people in the execution of specific tasks which have been identified as 'critical'.

5.2.1   *A conceptual view of behaviour observations*
Whereas all kinds of performance measurements have their value, the measurement of 'behaviour' is the most important one since this reflects the attitude of both the organization and the individuals working in it. It is the individual attitude of some people, namely top management, that serves as the foundation for the organizational behaviour which in turn influences the individual behaviour of others, and so on.

Individual and organizational behaviour are interrelated and they influence each other (see Figure 5.7, page 84). Through focusing on behaviour it will be possible to influence the attitudes of people. Therefore, the behaviour of people — both individually and as a group (the 'organizational behaviour') — is the most important measurement that can be made and is at the very heart of a safety management system. Everything people do originates from their attitudes and their beliefs. It is the behaviour that can be observed and measured, and which is a window on the attitudes behind it.

Desired behaviour is not something that happens by accident. It must be managed for best results. While doing that it must be kept in mind that 'desired' behaviour always refers to a standard, either written or otherwise. Any measurement must therefore start with determining that desired behaviour.

The 'management' of behaviour includes at least the following issues:

● The selection of people who already show the desired behaviour or who have the willingness and capability to show it. This selection becomes even more important nowadays as many businesses are moving towards an 'empowerment' culture.
● The way in which rules, procedures and so on are being set up in the organization. Such rules and procedures indicate the desired behaviour, even though they may not be written down. DNV's '24 points for obtaining compliance with rules etc' apply here (see Table 5.4). Specific qualities are demanded from those who make business systems (including hardware design) to have these systems operate with as few rules and procedures as possible. Necessary rules and procedures must be as close to 'normal' behaviour as possible. Both positive and negative factors influencing 'activators' and 'consequences' must be considered thoroughly so that undesired behaviour is discouraged and desired behaviour stimulated.
● The observation of the behaviour of individual managers and supervisors as well as workers to evaluate actual practice.

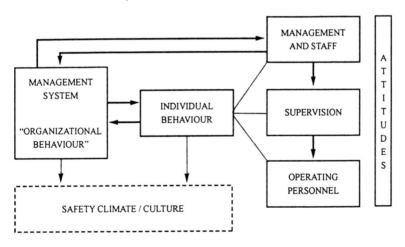

Figure 5.7   *Organizational behaviour, individual behaviour and attitudes*

In particular the observation of people in operating positions could lead to undesired situations in which people feel they are being 'spied upon'. So it is important to spend considerable time and effort on the introduction of such a programme. Furthermore, the behaviour of operating personnel is often the result of management behaviour.

Any worker observation programme should be well-planned and properly organized. The project organization must include higher levels of management to allow for appropriate decision making (and actions) in response to information gained through the worker observation programme. Eventually, observation activities must lead to modifications in the management system so that undesired behaviour is eliminated as far as possible from the organization.

A good way to start a worker observation programme is to run workshops with operating personnel during which the discussion includes:

● What is desired/undesired behaviour?
● What are the reasons for showing undesired behaviour?
● What works in safety and what does not?

The results of such workshops often point to issues which exist at higher managerial levels and need to be resolved there.

Table 5.4 *24 Points for 'How to obtain compliance with rules' or 'When to apply disciplinary action'*

PREPARATION	
1.	Rules only when necessary
2.	Involve people concerned in making of rules
3.	Explain why rules are necessary
4.	Make rules as simple as possible
5.	Rules must be justified and correct
6.	Make rules to comply with 'normal' human behaviour
7.	Rules must not contradict each other
8.	Make rule compliance attractive, make non-compliance unattractive
PRESENTATION	
9.	Give worker preview of rules prior to instruction
10.	Provide proper instruction
11.	Test knowledge of rules
12.	Keep rules accessible to users
APPLICATION	
13.	Provide 'try-out' period to test rules in practice
14.	Be clear in daily application of rules
15.	Provide sufficient time for proper rule application
16.	Be positive about rule compliance
17.	Be negative about non-compliance
18.	Proper example by supervision and management
19.	Maintain rule knowledge
20.	Periodically evaluate rule compliance
21.	Periodically evaluate effect of rules
22.	Keep rules up-dated
23.	Be clear about consequences in case of non-compliance
24.	Be consequent in case of non-compliance

Another aspect of these worker observations is that they should never replace day-to-day observations by line management. It is a first responsibility of line supervisors and managers to observe the people working in their area of responsibility in order to identify and correct behaviour that is not desired.

5.2.2   People in a leadership position
The evaluation of 'leadership behaviour' cannot normally be done through visual observation but requires interviewing the people concerned and, as necessary, people at the 'receiving end'. These evaluations are done using a set of criteria ('questions') for leadership performance. Table 5.5 shows example questions picked at random from a more extensive questionnaire. Such references are adapted to the function and provide the opportunity for scoring. Thus, quantitative indicators can be obtained.

Individual behaviour of people in a line management position is also reflected in the behaviour of the people working under their responsibility.

Table 5.5   *Examples of questions for leadership performance*

EVALUATION ITEM	MAXIMUM SCORE	ACTUAL SCORE
2. Appointed health, safety and environment (HSE) coordinator in his area		
9. Established HSE objectives and targets for his area		
11. Requires and supports HSE training for all managers/supervisors in his area		
16. Designated emergency coordinator in his area		
27. Makes HSE tours in his area at established frequency		
29. Takes part in annual HSE audits		
32. Sets leadership examples regarding rules compliance and adherence to safe work practices		
36. Chairs periodic HSE campaigns		
40. Discusses HSE policy issues with subordinates		
44. Includes HSE on agenda of all regular scheduled management meetings		
Total score	**Maximum 100**	**Actual 65**

Analysis
The quantified results of the 'observations' — that is, the interviews — can be used in different ways:

● for comparison with the maximum obtainable score (such as 65 out of 100);
● for comparison with the results from previous interviews of the same individual (for example 65 vs 55);
● for comparison with the average number obtained from a group of people in similar positions (for example 65 vs 75).

Table 5.5 shows an example of a scoring mechanism. Whereas the outcome of individual observations is certainly relevant, trend developments are more important to reveal any changes over a period of time.

Follow-up
As follow-up after the observations, the results are discussed with the individual and an action plan agreed upon to improve the individual's behaviour. To help in motivation, the reasons for showing desired behaviour are explained.

5.2.3   People in operating positions — general behaviour observation
Measuring behaviour of people in operating positions is normally done through visual observation.

General behaviour is concerned with the way people behave in a more general way while carrying out their job. The timing of general behaviour observations is normally not known in advance and is therefore 'unexpected' for the target group. Normally, these observations are concerned with groups of people working together in departments, rather than with individual employees.

General behaviour observations are made using a set of criteria for desired behaviour, to allow accentuation of the positive (see Table 5.6, page 88 — sample only). The set of 'behavioural aspects' can be obtained from the workshops which are organized for the introduction of the programme. The result of the observations is a percentage indicating observations with desired performance against total number of observations.

A risk factor can be obtained by applying a consequence factor which indicates the potential consequences of the observed undesired behaviour (see Table 5.6, last two columns). The risk factor is calculated by multiplying the number of undesired observations with the associated consequence factor. The latter is an estimation of the likely results should the situation remain unchanged. The consequence factor is very simplified here, ranging from 0.3 to 3. Any other method will do. The purpose of the exercise is to come up with a number which indicates the total 'risk level' resulting from the observation.

The various risk factors can then be totalled. The result may be an indication that follow-up action is needed. Since the number of undesired observations also depends on the time spent on the observation (the more time is spent, up to a certain point, the more undesired issues may be uncovered), it may be necessary to take into account the duration of the observation.

Acceptable and unacceptable levels need to be set per area and on the basis of experience. This should be done involving the work force. Those levels may change over time.

The risk level, and therefore also whether the level is acceptable or not, depends on the quality of the observation team. Inexperienced people will not recognize hazards even though they may be observing for a long time. Thus, using inexperienced teams will result in lower risk levels.

Table 5.6   *Form for evaluating general behaviour of people in operating positions*

BEHAVIOURAL ASPECT	TOTAL OBSERVATIONS	OF WHICH DESIRED	RATIO DESIRED TO TOTAL	CONSEQUENCE FACTOR	RISK FACTOR
Use of personal				3	
protective equipment				1	
				1/3	
Lifting				3	
				1	
				1/3	
Following				3	
procedures				1	
				1/3	
Working at proper				3	
speed				1	
				1/3	
Handling of				3	
chemicals				1	
				1/3	
Use of ladders				3	
				1	
				1/3	
Total			**80/100**		

Risk levels are relative. As an example, a particular company indicated that a level of 10 based on a one-hour observation (and concerning a certain area) was unacceptable and would require immediate management action. Between 5 and 10 was still unacceptable and in need for action. A score below 5 was considered acceptable (on average, while individual issues might still need attention).

Analysis
The quantified results of the observations can be used for comparison with:

● the maximum obtainable score (such as 80 out of 100);
● the results from previous observations within the same group or department (for example 80 vs 72);
● for comparison with the (average) results obtained from observing similar groups or departments (for example, 80 vs an average of 75).

Table 5.5 (page 86) shows an example of a scoring mechanism. Here also, while the outcome of individual observations is relevant, trend developments are more important to indicate any changes over a period of time.

Follow-up

As follow-up after the observations, the results are fed back to the group or department involved, and reasons for showing undesired behaviour are discussed. An action plan is set up with the group in order to improve its behaviour. Since the group's behaviour may be influenced by management, the action plan may also include discussions with people at higher levels in the organization.

5.2.4   *People in operating positions — task observation*

Task observations are directed at the performance of individual people in the execution of specific tasks which have been identified as 'critical'. The observations are done using a 'task procedure' indicating the critical steps of the task to be done and any steps to control undesired events.

The purpose of this type of observation is to evaluate the effectiveness of training programmes, to establish that procedures can still be executed as intended and possibly to improve the work procedures involved.

Carrying out task observations results in a report which indicates both positive and negative issues, and also actions to correct any deficiencies noted. Whereas these task observations would normally not result in a performance 'measurement', it would be possible to allow such a quantified evaluation.

Analysis

The outcome of a task observation is analyzed in relation to the effectiveness of training provided to the individual, the need for further training, the appropriateness of the work environment and the tools, materials and any changes that may have occurred since previous observations and which may necessitate a change of the task procedure.

Follow-up

The results of a task observation would normally be directed at the individual observed, in particular where it concerns individual aspects that need attention such as additional training. Depending on the outcome, however, the actions may also be directed at the group of people carrying out the task observed, especially when a common issue is discovered, including the need for updating the task procedure involved.

5.3   *Safety culture*
C Harrison, AEA Technology

The series of accidents in the 1980s turned the focus of regulatory bodies, safety practitioners and researchers away from the notion that accidents are caused solely through the actions of individuals. The accidents included the Chernobyl disaster, the Clapham (London) rail crash and the sinking of the Herald of Free Enterprise roll-on roll-off ferry off Zeebrugge. Emphasis moved towards management and organizational causes of disasters. In particular, since the mid-1980s there had been an increasing focus on the concept of 'safety culture'.

This concept was promoted by the International Atomic Energy Agency's International Nuclear Safety Advisory Group following the analysis of the Chernobyl accident[1]. It is being used increasingly to describe the overall attitude to safety within an organisation.

AEA Technology believes that a positive safety culture is one in which all employees — from top level management to individual workers — are committed to working safely and contributing positively to the safety of themselves and of their colleagues. The development of a safety culture within an organization is an iterative process; it becomes a corporate as well as an individual philosophy, and reflects strongly — at both a corporate and individual level — the commitment to establishing and continuously improving a safe working environment.

A positive safety culture is likely to be found in an organization described by Westrum[2] as 'generative'. Such an organization sets safety targets beyond ordinary expectations, emphasises results more than methods and values 'substance' more than 'form'.

Hence, safety culture within an organization is brought about by a combination of the prevailing values (sets of beliefs and attitudes held in common by workers and management), roles, social behaviours and work practices. Consequently the safety culture is influenced not only by factors which are directly related to safety but also to the general management structure and to the day-to-day work within the organization.

5.3.1   The importance of a positive culture

Culture is something that pervades a whole organization. It is intrinsic to the way individuals and managers respond and behave within a corporate framework. It is manifest in the frame of mind in which personnel undertake their tasks and responsibilities and the importance they attach to achieving overall company objectives.

A company that builds a positive safety culture encourages individuals and teams at all levels in an organization to be responsive and proactive in, for example:

● anticipating and managing risks and threats to corporate success;
● developing enhanced working practices and procedures;
● continuous improvement of operations, efficiency and the 'product'.

5.3.2   AEA Technology's approach to safety culture

In 1993, AEA Technology funded internally a two-year research project into safety culture. It followed the apparent organizational failings in many industrial disasters. At the end of this two-year period a comprehensive assessment methodology named the 'safety culture assessment tool' had been developed[3-5]. The purpose of this tool was to give organisations a means of identifying the status of their own safety culture.

The tool was developed as the result of a wide-ranging technical review. This review included not only the examination of contemporary work on safety culture but also areas such as organizational and management factors, team aspects, individual attitudes and responses, and factors influencing cultural change.

Another important part of this review was to look at the different types of studies that had been performed in industrial organizations. These included case histories of safety improvement schemes, studies which aimed to determine those factors which correlated with high levels of safety, and initiatives which claimed to improve not only safety but also safety culture itself.

This review resulted in an identification of 129 elements which were common to a positive safety culture. These became known as the 'safety culture parameters'. After closer examination, it became apparent that the parameters could be organized in one of nine key sub-groups and ultimately into three overall groups. This resulted in a 'safety culture framework' organized hierarchically (see Table 5.7).

Table 5.7   *The safety culture framework*

MANAGEMENT AND ORGANIZATIONAL FACTORS	ENABLING ACTIVITIES	INDIVIDUAL FACTORS
Positive organizational attributes	Reinforcement and incentives	Individual ownership
Management commitment to safety	Communication	Individual perceptions
Strategic flexibility	Training	
Participation and empowerment		
	SAFETY CULTURE PARAMETERS	

5.3.3   *Main areas of the safety culture framework*

It is at the management and organizational level that policies and procedures are set that provide the basic requirement for a positive safety culture. Hence, amongst other matters an organization should:

- demonstrate that it enforces practices that indicate a concern and commitment to safety;
- balance productivity with safety needs;
- set safety targets over and above that required by legislation;
- enable the workforce to share in the responsibility for safety;
- demonstrate that it can learn from past mistakes and be willing to learn and adapt.

Enabling activities are vital to any organization in order to improve and maintain a positive safety culture. They can be thought of as the means by which management initiatives are transferred to the workforce and also the means by which the workforce 'feed-back' to management. They include the provision of reinforcement and incentives, training (related to safety as well as work) and communication which is 'multi-directional' - that is, flows in all directions between management and the workforce.

Individual factors demonstrate that the individual's adoption of a personal responsibility for, or commitment to, safety is just as important to maintaining a positive safety culture as management's commitment to safety. In addition to that, the individual's perception of management's commitment to safety also has a bearing on safety culture, because it is this perception that ultimately influences the individual's safety performance.

5.3.4   The safety culture assessment tool
The three main areas within the safety culture framework provided the structure around which the safety culture assessment tool was devised. Each one of the areas is covered by an assessment methodology which allows a comprehensive examination of a company's safety culture. The management and organizational factors are assessed by a series of management interviews, the enabling factors by interviews designed for those responsible for training, safety and personnel issues, and the individual factors by a questionnaire completed by everyone in an organization. Interviews provide a qualitative assessment of the safety culture of an organization, whilst the results from the questionnaire are quantifiable.

The components of these interviews and questionnaire are based on the 129 safety culture parameters.

Management interviews
Management interviews are tailored to the role of the individual within the organization and are aimed specifically at staff with management responsibilities (senior, middle or supervisory). The overall aim of these interviews is to check that managers are aware of factors under their direct control that have an implication for a positive safety culture and have put systems into place to control them. The interviews also allow a comparison of how different levels of management impact upon the organization's safety culture. The questions that make up the interviews allow the interviewees to make 'open ended' responses. Examples of the types of questions interviewees would be asked and the sort of responses required are shown in Table 5.8.

Table 5.8 *Management interview: examples of questions and ideal responses*

EXAMPLE OF QUESTION	IDEAL RESPONSE
How often do you discuss safety issues?	The respondent would have to demonstrate that safety issues are regularly discussed with representatives from every level of the organization
What is the relationship between management and regulators?	The respondent would have to demonstrate that the relationship was open and cooperative
How important is team work?	The respondent would have to demonstrate a positive attitude towards teamwork and evidence of it in the organization
Are individuals able to voice concerns over safety without fear of reprisals?	The respondent would have to demonstrate that the organization was one with a blame-free culture
How often do you intentionally visit the shopfloor?	The respondent should indicate that regular visits are made to the shopfloor demonstrating that management is visible to the workforce

Enabling factors interviews
Enabling factors interviews have been designed specifically to determine whether sufficient and suitable initiatives are in place to inform, train and motivate personnel, all of which have

implications for the company safety culture. Representatives of the training department are interviewed to gather information about safety and skill training, and those responsible for human resources interviewed to gather information on disciplinary procedures, incentive schemes and the systems that are in place to convey information to the workforce. Finally, those with safety-related functions are interviewed to obtain information on the way in which safety-related information is exchanged and relayed throughout the organization. The majority of the questions that make up the interviews require 'yes/no' answers. Examples of the questions asked in an enabling factors interview and the ideal response required are shown in Table 5.9.

Individual questionnaire

The individual questionnaire is given to all personnel within an organization. Its purpose is to determine individuals' attitudes, opinions and perceptions of safety. The questions also cover issues which influence morale and conditions on the site which can cause unnecessary stress, and which in turn can lead to unsafe actions. It also examines individuals' perceptions of management actions and abilities to provide a safe working environment. Perceptions are important because they influence actions. Mismatches between the situation as it really is and an individual's perception of that situation influence safety culture.

Table 5.9 *Enabling factors interview: examples of questions and ideal responses*

EXAMPLE OF QUESTION	IDEAL RESPONSE
Is basic safety training provided for new recruits?	Yes
Is the performance of trainees assessed?	Yes
Are the contents of the safety policy made available to everyone?	Yes
Is information from accident and incident investigations fed back to the workers?	Yes
Are awards given for safe behaviour?	Yes
Are disciplinary measures taken when safety rules are wilfully ignored?	Yes

The majority of the questions require individuals to choose the option ('strongly agree/agree/disagree/strongly disagree') which best fits their response to the question. The ideal response in all questions is to choose 'strongly agree'. Examples of questions from the individual questionnaire are shown in Table 5.10 (see page 94).

All three of these methodologies should be used to provide an overall picture of the organization's safety culture. The results of the assessment highlight an organization's strengths and weaknesses. Recommendations can then be made about how to build on strengths and overcome weaknesses. Once these measures have been put in place and have had a chance to consolidate within an organization, it would be prudent to re-assess the safety culture of the organization to see if it has improved.

5.3.5   Case studies

As well as being applied 'in-house', AEA Technology's safety culture assessment methodology has been applied within a number of external companies. These companies represent the nuclear sector (both in the UK and in Eastern Europe), the rail sector and the oil and gas sector. The case studies which follow provide an overview of the work done.

The UK nuclear sector

The safety culture assessment tool was applied within two divisions of the same company (commissioned as two separate studies). Both studies were in response to the UK Health and Safety Commission's Advisory Committee on the Safety of Nuclear Installations report on human factors, 'Organizing for safety'. One of the recommendations of this HSC report was the need to encourage and develop a positive safety culture. To ascertain the most appropriate ways in which to do this, it was first necessary to establish the level of safety culture within the organization.

Table 5.10   *Individual questionnaire: examples of questions*

EXAMPLES OF QUESTIONS
Safety issues are given priority by my senior managers
My company provides plenty of safety information for all members of the workforce
Safety is my responsibility
My company is committed to training staff at all levels
Morale is good in my department/section
Senior management will acknowledge a good safety record

In the first study, 26 interviews (management interviews and enabling factors interviews) were conducted with a cross-section of staff from all areas of the organization including representatives of senior management, shift managers, health physics, contractor liaison, training, personnel, area safety co-ordinators and trade union representatives. Individual questionnaires were distributed to all 1100 staff (571 were returned).

The results from the interviews were combined with those from the questionnaire. In qualitative terms, the safety culture was identified as being fairly positive but in definite need of improvement. In quantitative terms, the mean subject score on the individual questionnaire was 241 (out of a possible 370), confirming the qualitative evaluation. Particular problems that were identified were:

● Lack of communication of both safety-related and general information.
● No schemes to encourage and promote awareness of safety.
● Lack of active participation in decision making: individuals showed a lack of initiative.
● Conflicting messages concerning the balance of productivity with safety: the verbal message did not correspond with people's experience.
● Management's commitment to safety and the workforce was not always apparent.

Recommendations were offered to help tackle these problems. For instance, in light of 'lack of communication of both safety-related and general information', the division was advised to implement 'team briefings' across the division and ensure that those people working shifts participated in the system by holding meetings once a week. They were advised that the teams should not consist of more than 12 people, that teams should comprise a cross-section of staff with a team leader, and that checks should be made to ensure that these meetings actually took place. Safety should always be on the agenda of these meetings. This keeps the focus of activities on safety and provides people with a valid reason for raising safety issues.

In the second study, 19 interviews were conducted with a similar cross-section of staff as in the first study. 2069 questionnaires were administered to staff with a return of 965. The results from both the interviews and questionnaires suggested that the safety culture was fairly positive. Particular problems that were identified were:

● There was a haphazard approach to risk assessment: no formal risk assessment for certain operations; staff were left to assess risks from day to day.
● Formal investigations were not always conducted following incidents.
● Control of contractors was poor.
● Enforcement of disciplinary procedures when an individual intentionally ignored safety procedures was not particularly good.
● Points of contact with certain regulators were not as good as they should have been.

Recommendations were offered to combat these and other problems that were identified.

Once these two studies had been completed, the results from both were combined to give an 'overall' impression of the safety culture. It was found that, despite different strengths and weaknesses between the two sites, there were some common problems:

● Communication (of issues related to both safety and general work).
● Raising safety awareness.
● Control of contractors.
● Relationships with regulators.

It was also demonstrated that personnel who worked at the division from the first study were more prepared to accept a broader definition of safety than those who worked within the division from the second study. In other words, they were more ready to accept that 'safety culture' embraces issues that, at face value, do not appear to be relevant to safety, such as morale of the workforce, training and incentives for safe working.

The Eastern European nuclear sector
It had been recommended to the organization involved in this study that they should improve both their safety policy and safety culture. Hence, AEA Technology were asked to develop a methodology and plant-specific guidance for regular self-assessment of the achieved safety culture level within the organization. The organization required the methodology to include not only a safety culture assessment tool, but also two sub-sets of questions for inspection schedules and incident/accident investigations schedules. This methodology would allow for

the results for all three methodologies to be combined, leading to more informed measurement of the safety culture of the plant.

The first task was to develop a master set of questions. The management and enabling factors interviews and the individual questionnaire were used as a basis for this master question set. After consultation with safety personnel during an on-site visit, the questions were agreed. They were organized into eight 'key groups'. Within each of the key groups, the questions were organized into one of two formats: a question format ('strongly agree-strongly disagree' response) or an interview format (open-ended response). It was advised that a comprehensive assessment of safety culture would require a combination of the formats. In addition, some questions could only be asked at the management level and were designated as such.

Within each of the eight key groups, the questions were organized into subject areas. Each of these subject areas was assigned 'indicators', which described those responses to questions that indicated a positive safety culture.

The inspection and incident/accident schedules and the main safety culture assessment were then developed by extracting applicable questions from the master set of questions. It was understood that questions could be added to the main safety culture assessment tool from the master where and when this was appropriate.

A processing model was then developed which demonstrated when assessments should be conducted, how they should be conducted and how the collated data should be analyzed.

Finally, staff from AEA Technology conducted an on-site training course in the use of this tool for safety personnel from the organization.

The rail sector

In order to determine the effectiveness of established safety systems across a particular railway group, the need to assess safety culture had been highlighted. The group recognized that the issue of safety culture was inadequately covered by its existing audit systems. As a consequence, AEA Technology was commissioned to develop a methodology for assessing safety culture in the context of the railway industry. However, AEA Technology was not required to administer the tool once developed. This would be done by personnel who worked within this railway group.

The railway group had decided that it would like to administer a number of different types of safety culture assessments depending on the particular needs of the group at the time. These fell into the following categories:

● Overall assessment of safety culture (management level).
● Overall assessment of safety culture (other worker level).
● Detailed assessment of key groups.

AEA Technology devised a methodology that would accommodate all three types of assessments. Using the management interviews, enabling factors interviews and individual questionnaire as a basis, AEA developed 'master lists' of questions (with the approval of the rail group personnel). These master lists corresponded to the nine key groups. Within each of these lists, some of the questions were designated as core questions considered to be essential for any of the assessments, with a series of detail questions which could be used to supplement the core questions where appropriate. In addition, others were identified as management questions and answers were only expected from those in a management role. Hence, an 'overall assessment of safety culture' at the 'other worker level' for instance, would have to include all the core questions across the nine master lists but none of the designated management questions.

This methodology was then transferred to software. The software tool consisted of three utilities:

● A question maintainer which contains the master sets of questions as described previously, modified to suit the needs of the user.
● The audit tool which contains a sub-set of questions to be used within a particular audit. They are presented to the individuals being audited via a simple user interface containing the chosen questions set and a simple choice of responses.
● An analysis module which combines data from one or more audits with the same base set of questions, and presents the data in tabular or graphical format.

This software is currently with the railway group which is about to embark on a pilot study.

The oil and gas sector

A petrochemical company wanted to gather an indication of the level of safety culture at the 'worker level' on a number of their offshore installations. Employees were asked to complete a version of the individual questionnaire that had been adapted for specific use within the petrochemical industry.

Out of the 900 personnel approached to complete the questionnaire, 770 were returned. In general, there was positive feedback about the level of safety culture within the three installations. There were some areas where improvements needed to be made, however, to encourage a more positive safety culture. Some of the main problems are shown in Table 5.11.

Table 5.11 *Offshore case study: selection of main problems*

INSTALLATION 1	INSTALLATION 2	INSTALLATION 3
● Problems getting hold of personal protective equipment (PPE)	● Employees felt disciplinary procedures were unfair	● Accommodation areas were poor
● Personnel forgot safety when they were away from the job	● Good safety performance went unrewarded	● PPE not available
● Work equipment not adequate	● Incident reporting not done	● Training needs not addressed
● Problems with environmental issues - eg: noise levels		● Getting the job done was often placed above safety
		● Employees not consulted with regards to safety issues

5.3.6 Conclusion

Section 5.3 has addressed a number of issues. It first explained why the concept of safety culture has emerged and why it is so important. It then discussed AEA Technology's understanding of the area of safety culture and how the company developed a methodology to assess the level of safety culture within an organization. Finally, it went on to discuss the work that AEA Technology has done in the field of safety culture.

What is evident from this work is that the circumstances that determine a safety culture assessment differ between organizations. AEA Technology's safety culture assessment methodology was developed with this in mind and has been easily adapted to suit the different requirements of different organizations. These needs have included the introduction of industry- and culture-specific questions, a 'methodology for self-assessment' and the transfer of the assessment methodology into software.

Lessons learnt

AEA Technology has found that people's understanding of safety and safety culture affects the way in which they respond to an individual questionnaire. It was not uncommon for people to not answer questions that they felt 'had nothing to do with safety'. Some people seem to have difficulty understanding that safety culture embraces issues such as training, morale, communication, incentives and perceptions. Hence, educating the workforce is important if a safety culture assessment — and actions taken as a result — are to have any chance of making a positive impact on an organization.

AEA Technology's work in Eastern Europe has also demonstrated that not only are there differences between different organizations in their understanding of safety culture but that these differences also exist at a cultural level.

5.3.7 References

1. IAEA, 1991, *Safety Culture: A report by the International Nuclear Safety Advisory Group*, Safety Series No 75-INSAG 41991.

2. Westrum R, 1988, *Organizational and inter-organizational thought*, Proceedings of World Bank Workshop on Safety Control and Risk Management (Washington D C, USA).

3. Bone C, Falconer H and Lucas D, 1993, *Safety culture management report*, AEA/CS/40245001/Z/2.1 CIRE Project 93-7003 93/02.

4. Bone C, Falconer H, and Lucas D, 1994, *Safety culture framework*, AEA/CS/-402450001/Z/4.1 CIRE Project 93-7003 93/03.

5. Bone C, Harrison C and Lucas D, 1994, *Validating the safety culture parameters*, AEA/CS/40245001/Z/6.1 CIRE Project 93-7003 93/04.

Note that other references used in the writing of this section could not be detailed here because they were reports written for commercial clients of AEA Technology plc.

5.4   *Tripod incident analysis methodology*
P G D van der Want, Shell International Chemicals

Tripod incident analysis methodology aims to identify within organizations the flaws that play a vital role in causing incidents; it categorizes them to enable structured remedial action. It has been based on research work carried out by both Leiden University and the University of Manchester.

5.4.1   *Incident causation analysis methodology*
An incident is the product of a long chain of events. In the conventional view of how incidents happen, a number of unfortunate events come together and form an incident. This view usually includes a specific human component (unsafe acts) which means that the incident would not have happened if the unsafe acts had not been performed. As a consequence, there is a widespread tendency in all organizations to blame incidents on the people who suffer them. Incident investigations frequently cite human error as the cause and many people still see such errors as the beginning and end of the incident sequence.

Another important element in this view of incident causation is the breach of defences. Defence systems are the back stops intended to protect against unlikely but possible circumstances, such as excess flow valves, high-temperature alarms, protective clothing or emergency procedures. They are triggered when other failures have occurred. Because they are used infrequently, they can be neglected and may fail, thus providing the window of opportunity without which an incident would not happen or would be limited to a non-damaging event.

The breached defences and unsafe acts are tangible, well understood aspects of incidents. They emerge relatively easily from the fact finding and are confined to the immediate time period and surroundings of the incident. They are called active failures, because they are physically measurable and the time period for failure to manifest its adverse effect is short.

The Tripod concept
In the Tripod concept, incidents are perceived as the final chapter of a much longer story. Active failures do not occur in isolation but are encouraged by existing preconditions. These are the conditions underlying many human errors. They depend on the person as well as on the circumstances, and are often hard to detect and to quantify. Examples are inattention, haste, stress, misperception, competing demands, ignorance, poor motivations or the time of the day.

Preconditions are caused by latent failures which could originate much earlier in the organization and from decisions of line management. Line management activities are removed in time and place from the end of the line operations, where the incidents occur. Therefore the adverse consequences of these so-called latent failures may lie dormant within the system for a long time, only becoming evident when they combine with other factors to breach the system defences.

Behind the latent failures it may be possible to identify the fallible decisions which initiated them. These decisions establish company strategies and policies which are the principal organizational sources of latent failures, failures which could not have been foreseen at the time but often manifest themselves years later.

The incident itself, active failures and latent failures are the three 'feet' in incident causation that give the Tripod framework its name. The sequence of elements in the incident causation chain and the segment of the organization where these elements are most likely to occur are represented in Figure 5.8.

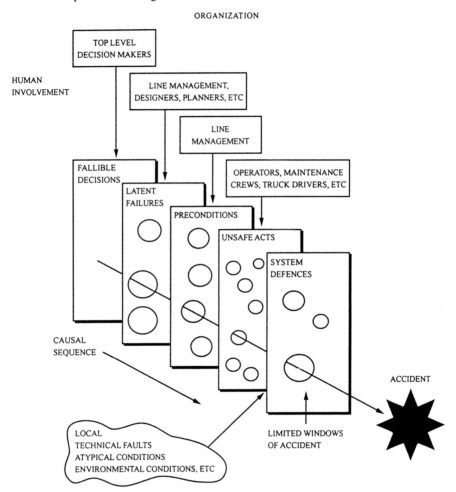

Figure 5.8   *Accident causation sequence*

5.4.2   *General failure types*

Based on in-depth analysis of incidents in various parts of the Shell organization, the Tripod methodology categorizes latent failures in 11 general failure types (GFTs) that are applicable to Shell's type of business. In order to understand how these GFTs are capable of building up a broad picture of safety, it is necessary to appreciate how each one relates to individuals as they carry out their day-to-day activities in the workplace. This shows how the GFTs relate to each other, and has a close link with the incident causation sequence.

The individual is at the centre of the Tripod web (Figure 5.9). The GFTs vary in their proximity to the individual but they can all lead to an incident.

The closest GFT to the individual is defences (1). These measures are specifically designed to mitigate the consequences of either human or component failure once they have occurred. If an incident is caused by failures of other GFTs, then adequate defences can still save the day.

The next layer of the web is error-enforcing conditions (2). This is the GFT which is most closely connected in time to the commission of unsafe acts by individual members of the workforce. For example, shift change could occur close to the time of an accident, while the organizational decisions which have led to the timing of shift changes might have taken place some time before. Other examples are the physical working conditions (hot, cold, noisy, etc), acting on the individual or on the workplace and promoting the performance of unsafe acts in the form of errors or violations. An error-enforcing condition almost always implies another GFT. For example, unfamiliarity — a condition in which a person has to think, and make critical decisions with little or no experience — implies the GFTs 'training' and/or 'procedures'.

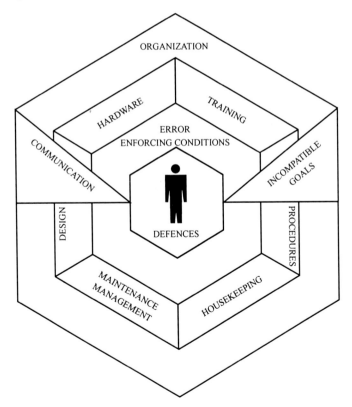

Figure 5.9 *General failure types (GFTs)*

Training (3) includes the deficiencies in the system for providing the necessary awareness, knowledge or skill to individuals in the organization. In this context, training includes the job coaching by mentors and supervisors as well as formal courses. Procedures (4) concern unclear, unavailable, incorrect or otherwise unusable standardized task information that has been established to achieve a desired result.

Error-enforcing conditions receive direct influences from several other GFTs: 'defences', which were discussed already, plus 'design', 'hardware', 'maintenance management', and 'housekeeping'.

Design (5) includes deficiencies in layout or design of facilities, plant, equipment or tools that lead to misuse or unsafe acts, increasing the chance of particular types of errors and violations. Hardware (6) includes failures due to inadequate quality of materials or construction, non-availability of hardware and failures due to ageing (position in the life-cycle). Maintenance management (7) is about failures in the system for ensuring technical integrity of facilities, plant, equipment or tools - for example, condition surveys, corrosion controls, and function testing of safety and emergency equipment. Housekeeping (8) captures the tolerance of deficiencies in conditions of tidiness and cleanness of facilities and work spaces, or in the provision of adequate resources for cleaning and waste removal.

These six additional GFTs (3-8) form the next layer of the web in Figure 5.9. They are all closely connected with day-to-day operational matters. These in turn are shaped by the three remaining GFTs: 'communication', 'organization' and 'incompatible goals'.

Communication (9) includes failure in transmitting information that is necessary for the safe and effective functioning of the organization to the appropriate recipients in a clear, unambiguous or intelligible form. Organization (10) describes the deficiencies in either the structure of a company or the way it conducts its business that allow safety responsibilities to become ill-defined and warning signs to be overlooked. Incompatible goals (11) is about the failure to manage conflict: between organizational goals such as safety and production, between formal rules such as the company's written procedures and the rules generated informally by the work group, or between the demands of individuals' tasks and their personal preoccupations and distractions.

These last three GFTs (9-11) are known as the higher order GFTs. Organization is furthest away from the individual at the workplace but, nevertheless, failure in this area can lead to incidents. Failures in the other two can lead to accidents when triggered by specific local circumstances. These can manifest themselves across all the areas covered by the other GFTs. Hence, they cut across the web rather than being another layer.

Here is an example to illustrate the overall concept. A gas leak occurs unnoticed because the gas detection system has not been properly maintained and serviced. The active failure in this case is the failure to detect the gas leak. This is caused by the precondition that the gas detection system is allowed to remain unmaintained and unserviced. The latent failures in the system are the reason why this has happened. The incident could be attributable to:

● design, because there was no way to test the system;
● incompatible goals, because the system was designed to be tested on shut-down but shutdowns were not allowed;

● organization, because the procurement system did not allow for the supply of test gases;
● maintenance management, because the management system did not call for testing the gas detection system.

5.4.3   Failure state profiles

Another way of looking at GFTs is as a means of predicting failure before the event. One of the problems people face when they make fallible decisions is lack of knowledge of the bad consequences. After an incident it may be relatively easy to trace back and see how decisions led to problems, but hindsight is always perfect. Tripod is aimed at proactive measures, and thus at preventing the next accident. Using the GFTs offers a structured way of predicting the likelihood of problems cropping up in the future.

Tripod research has demonstrated that assessment of the degree to which these GFTs are present in an activity or facility provides an accurate picture of its overall 'safety health'. The assessment may be quantified in several ways. The relative presence or absence of each of the 11 GFTs may be represented by the height of a bar in a histogram format. This histogram is called a 'failure state profile' (see Figure 5.10). In this example, 'communication', 'organization', 'procedures' and 'incompatible goals' are the most problematic GFTs (the higher bars). Available resources should be concentrated on addressing these issues.

The two main applications of failure state profiling — that is, reactively in incident analysis and proactively in the 'Tripod Delta' technique — are described below.

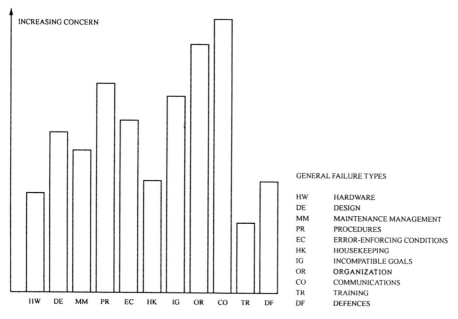

GENERAL FAILURE TYPES

HW	HARDWARE
DE	DESIGN
MM	MAINTENANCE MANAGEMENT
PR	PROCEDURES
EC	ERROR-ENFORCING CONDITIONS
HK	HOUSEKEEPING
IG	INCOMPATIBLE GOALS
OR	ORGANIZATION
CO	COMMUNICATIONS
TR	TRAINING
DF	DEFENCES

Figure 5.10   *Example of failure state profile*

5.4.4   Tripod in incident investigation and analysis

Tripod is not unique in identifying a causal sequence of events. Other incident causation theories in the literature distinguish 'direct causes' and 'underlying causes'. Its strength is in helping to translate from the active failure mode, of which evidence can be found near the incident (time and place), to the latent failures, which are more hidden and remote. This latter category — that is, underlying causes — is more fundamental for health, safety and environmental management. In practice, the results of the Tripod analysis are used to structure the findings of incident investigation at five levels of causal sequence. Subsequently, the findings are used to classify the underlying causes into GFTs which need to be remedied.

The various steps in the analysis process may be summarized as follows. All items of factual evidence are reviewed and classified broadly in one of the five Tripod main levels:

- Identify breached defences/barriers that failed or were missing.
- Identify unsafe acts/omissions and categorize human errors/violations.
- Determine the preconditions relevant to each unsafe act.
- Identify the latent failures and categorize these into GFTs.
- Identify the (fallible) decision behind the GFTs where possible.

This process is repeated and missing information is sought out. When applying the causal sequence, each chain of causes must be complete with linkage between all of the five levels of the sequence. An incident causation tree can then be prepared, and the GFTs identified can be presented as a failure state profile.

The results of several such analyses can be combined in a histogram to produce a failure state profile of an operation which clearly shows weaknesses in an operation by way of the most frequently involved GFTs.

5.4.5   *Diagnostic tool for accident prevention*

Fundamental to the Tripod theory is the idea that when a GFT is present in an organization it will be characterized by a number of conditions or 'token failures'. These conditions are recognizable, but tolerated or endured because of their perceived low significance. This has been recognized as an opportunity to develop a proactive application: to use the tokens in a diagnostic way to find latent failures and thereby to be able to eliminate them before they can contribute to an incident.

The ability to analyze an operation proactively, not just by retrospective analysis of incidents, is seen as a significant breakthrough in safety management. It came at a time in the Shell organization when a number of operating units, after having radically improved accident frequency rates, were looking for some way in which they could best direct their safety efforts to sustain performance, without the stimulus of accidents.

A tool has been developed called 'Tripod Delta', to distinguish it from the application of Tripod in incident analysis, and has been in use as an experimental tool since 1993.

The heart of Tripod Delta is a database of indicator questions that would return a 'fail' response when the condition failed to meet a predefined acceptance level. Normally this database contains approximately 2200 questions for a specific operation. Before application the basic database has to be customized and calibrated to the organization in which it will be used. Customization ensures that the questions are adapted to represent situations which are typical of the operations, and which will be recognized by the staff involved. In the calibration step, positive correlation of each question with a GFT is verified, and the level of difficulty of subsequent questionnaires is adjusted. This process ensures adequate repeatability of the Delta sampling process.

Next, a questionnaire is developed consisting of some 220 questions selected randomly by a software program, but evenly spread over the 11 GFTs. The workforce in the organization is then asked to complete the questionnaire. The answers are interpreted and translated into a failure state profile. This profile can be used in a way identical to the profile developed from incident analysis to address latent failures and achieve improvement.

Typical examples of Tripod Delta questions and their correlating GFTs are:

● 'Has overtime been required daily during the last week?' which correlates with 'error-enforcing conditions'.
● 'In the last month have you had a training course cancelled without advance notice?' which correlates with 'communication'.
● 'Are the short-term goals always passed on at shift changes?' which correlates with 'incompatible goals'.
● 'Is the most recent (operator) department organogram permanently displayed on a notice board?' which correlates with 'communication'.

After six months the same workforce may repeat this exercise with another set of indicator questions selected by the software program. This should confirm improvements or give a similar profile if no action had been taken.

The experimental status of the tool is related to certain constraints in its application. It is an appropriate tool for operational units where the incident frequency is very low. In these units, management is concerned with continuing to improve the performance level but does not have failures highlighted by incident investigation as indicators of where preventive effort should be concentrated. In an environment with high total recordable injury frequency, Tripod Delta is less appropriate.

It has also been observed that Tripod Delta profiles tend to become flatter, which is associated with lower reliability, when the tool is used on samples which are too coarse, like whole refineries or large chemical sites. The tool is seen at its best when profiling smaller operational units.

5.4.6   Implications and benefits
Tripod is not a tool which can be implemented easily. Incident analysis teams need to receive dedicated training, but also the workforce not directly involved in the analysis should obtain a basic understanding of the purpose of the tool and the way it will be implemented. Furthermore the culture in the organization must be such that exploring discussions into issues that at first hand seem not directly related to an incident are tolerated.

The calibration process in Tripod Delta incorporates information from eight to ten comparable units before the desired statistical reliability can be achieved. Thus, a substantial preparatory effort is required before the tool can be made operational.

The benefits of Tripod are, however, not only the insight into the real reasons why incidents happen and how an organization's resistance to incidents can be improved. It also provides feedback on potential incident causes at a much earlier stage in the causation sequence, and a way to prioritize scarce resources in the improvement process. Finally, it sets a more positive climate for feedback in safety, stressing accountability of the individual rather that the attribution of blame.

Section 5.4 is a compilation of abstracts from the 'Incident investigation and analysis guide' by Shell Internationale Petroleum Maatschappij BV, Den Haag, 1993, and from other Shell Tripod training documentation. Tripod and Tripod Delta are Shell trademarks.

5.5   *Process safety management performance measurements*
W J Kolk, DuPont

In the DuPont company, many safety performance measurements are used to assess plant and equipment, systems and procedures and people. Although these distinctions are recognized, the 'people' aspect is viewed as an integral part of all aspects of the business. Therefore, since the DuPont system of process safety management (PSM) encompasses a wider perspective on safety performance measurements, a brief explanation of this system and its relationship to management systems and behaviour is provided here.

5.5.1   *Four steps to effective PSM*
The following four key steps have been identified within DuPont as necessary in order to manage process safety:

● Establishing a safety culture.
● Providing management leadership and commitment.
● Implementing a comprehensive PSM program.
● Achieving operating excellence.

These steps have been presented in several corporate publications and at conferences[1, 2]. It is not the intention to elaborate on the steps here, but rather to highlight aspects of them in order to clarify their importance and the importance of safety performance measurements within the total safety system.

Establishing a safety culture — the first step
A PSM program — established and implemented in an organization with strong dedication and commitment to safety, health, and environmental issues — will flourish and continually be strengthened and reinforced by management decisions and actions. To be effective, the safety culture must be an integral part of an organization's beliefs and core values.

The DuPont safety culture is anchored by a formal corporate safety mission, philosophy and principles that provide guidance to employees when other business needs are competing for resources and priorities. The principle 'All injuries and occupational illnesses can be prevented' is the central belief in the DuPont system.

Promoting safe behaviour, both on and off the job, requires extensive programmes

Promoting safe behaviour, both on and off the job, requires extensive programmes and training and is an ongoing effort. The ultimate goal of this effort is employees who have a sense of 'ownership' for safety and the dedication to follow correct procedures each and every time, and who help their colleagues achieve the same state.

Comprehensive programmes are in place to audit and measure safe behaviour. These programmes are discussed in sub-section 5.5.2 on safety performance measurements.

Providing management leadership and commitment — the second step
Management leadership and visible commitment form the foundation of any lasting effort to achieve and sustain excellence in PSM. This is characterized by management actions that continually support and reinforce the company's goals and policies. It is important that these actions begin with the most senior management and extend down through each level in the organization.

Besides establishing a safety culture, management responsibilities important to PSM include:

● establishing PSM policies and guidelines, and committing resources necessary to implement them;
● involving employees;
● establishing clear accountability for performance against specific process safety goals and/or objectives;
● verifying, through measurements, the degree of compliance versus established PSM policies and guidelines;
● personally participating in activities that visibly demonstrate commitment to PSM.

It is important to stress the involvement of employees in the safety process. Employee involvement is specifically mentioned in the US Occupational Safety and Health Administration (OSHA) PSM regulation[3] and in the new EU Seveso II Directive[4]. In DuPont, it is also a strategic intent.

DuPont, like many other companies, is involved in change aiming to maximize the contribution of all employees, moving the focus from what people do (their function) to their motivation and approach (their role). The company believes these work systems enable an organization to make and sustain a stepchange in safety performance and, in fact, in every aspect of management.

Implementing a comprehensive PSM program — the third step
DuPont has introduced and implemented throughout the company a PSM system that consists of 14 elements, grouped around the following segments:

● Technology.
● Facilities.
● Personnel.

Each of these resource areas has several important elements, as shown in the model represented in Figure 5.11.

All seven elements for a safety management system, as specified in Annex III of the Seveso II Directive, are included in the DuPont system. The OSHA PSM regulation is also very similar to the long-established DuPont system.

The basic element of the DuPont system is 'documented technology' (process safety information), which requires that the hazards of the chemicals, the process design basis and the equipment design basis are known and are easily accessible throughout the life of a facility. Other elements, such as operating procedures, management of change and mechanical integrity build on that element. It is important to note that change is defined as any deviation from the documented technology.

The logic of the DuPont PSM system is as follows:

● Start with a safe plant. Before permission to operate the new facility is given, the following must be in place: technology known and documented, inherently safe design, quality assurance during fabrication and installation, trained personnel and operating procedures available. A pre-startup safety review must also be conducted.
● Preserve *status quo* after startup by maintaining mechanical integrity and appropriate training levels, and by learning from incidents.
● Manage change in the areas of technology, personnel and facilities.
● Audit for compliance and verify the effectiveness of the system.

A comprehensive system is in place at DuPont to measure and quantify this compliance. It is important to mention that the 'people aspects' are audited in all 14 elements of process safety.

Figure 5.11   *DuPont's process safety management model*

Achieving operating excellence — the fourth step
Operating excellence is used to describe an organization in which every member of that organization has developed a deeply-rooted dedication and commitment to carry out each task the right way every time. The safety culture is the foundation on which the system is built, and operating excellence and sustainable improvement are the results of implementing and sustaining that system.

5.5.2   *Safety perfomance measurement*
Safety performance measurements include both lagging and leading indicators. Lagging indicators (injury performance) are classified according to the OSHA system as:

- medical treatment cases;
- restricted workday cases;
- lost workday cases.

Lagging indicators are described in detail in section 6.1. However, for reference purposes here, the current injury performance of the top ten DuPont manufacturing sites is given in terms of safety performance in sub-section 5.5.6. The majority of injuries at DuPont manufacturing facilities are not the result of process-related incidents.

The most prominent leading indicators (observation programs) include:

- Unsafe acts index.
- PSM audit scoring system.
- Process incident scoring system.

5.5.3   *Safety performance measurements — unsafe acts index*
An unsafe acts index expresses results quantitatively from periodic observations of the behaviour of personnel. The index can be the number of observed unsafe acts or conditions per audit, per hour or per average number of employees present in a certain area. Unsafe acts can be split further into categories such as:

- use of personnel protective equipment;
- ergonomics;
- actions;
- conditions (behaviour rather than systems-related).

Observation programmes are carried out by trained personnel of their own work area, by personnel of other areas, by site management and by safety specialists. Calibration is obtained by using the results of one group — usually the safety specialists — as a benchmark. Ratings are not shared with other sites; they are purely for internal use, to recognize strong areas as well as areas requiring improvement.

The most common observation training program is STOP (safety training observation programme), a DuPont programme that is also widely used outside DuPont. The essence of the STOP technique is knowledge of human behaviour.

In these observation programmes at DuPont sites, there are trends toward:

● increased involvement of all levels, through observations conducted by personnel on the work floor;

● strong participation by contractors;

● improved interactive skills (talking with people) when deviations from safety standards are observed (stimulating thinking about safe behaviour, rather than criticizing);

● critical review of both the need for additional guidance or training and the usefulness of the existing rules and procedures.

Participation by all levels of the organization and by contractors is a strong motivator for developing safety consciousness.

5.5.4   Safety performance measurements — PSM audit scoring system

Operations at DuPont sites worldwide are audited periodically by teams from other DuPont locations. These audits, which normally take four days, consist of a review of the status of all 14 PSM elements, using an interactive process. Both systems and performance audits are conducted, as is the case with ISO 9000[5].

Comprehensive audit protocols and checklists are available. Checklist questions carry a certain value, based on their significance in the total system, with people-oriented matters having the highest weight. A percentage of the total score value for each question is determined, based on the audit findings.

This interactive process means that the auditor is carrying out an in-depth analysis. It requires trust and confidence from both the auditor and those being audited. Motivation for operations personnel to cooperate and be open stems from the fact that audits lead to recommendations that help those same personnel in their efforts to achieve continuous improvement, and also could prevent serious incidents. In other words, they gain from the auditing experience.

Qualified PSM auditors are DuPont personnel with a high level of operational experience and knowledge of the corporate PSM system. The formal corporate PSM training programme is very similar to ISO's internal training programs (auditing techniques).

PSM audit protocol

The objective of the audit is to:

● determine the effectiveness of the PSM system in meeting:
  - DuPont standards;
  - regulatory requirements;
  - site expectations;
● determine the effectiveness of the documentation systems;
● assess effectiveness of corrective actions;
● coach and counsel personnel regarding DuPont standards.

In order to do this effectively, a performance audit is broken into three sections:

● Preaudit questions — designed to evaluate whether requirements of DuPont guidelines are contained in the site's PSM procedures and guidelines.

● Verification questions — designed to evaluate the facility's compliance with PSM procedures and guidelines. This includes field-checking documents, records, manuals, etc.
● Interview questions — designed to evaluate the effectiveness of the PSM program implementation.

Audit results
Audit results are reported in a comprehensive document that highlights the pockets of excellence in an operation as well as areas requiring improvement. Recommendations are documented for each question. Site organizations, together with site management, then take corrective actions to ensure continuous improvement. Feedback on corrective actions is provided to business unit management, auditors and the corporate safety, health, and environment (SHE) excellence centre.

Audit scores are given for each individual element as well as for the total audit. This score is a corporate 'metric' and is reported to site and business management. (No comparisons are made between sites or business units.) Individual element scores provide input to corporate PSM technology element teams for development of additional assistance.

5.5.5   Safety performance measurements — incident scoring system (including near-misses)
A simple and quick checklist is used to rate process incidents. Table 5.12 is an example of the scoring system applied to each of ten checklist questions.

Table 5.12   *Example of a scoring system question*

SIZE OF RELEASE OR FIRE	POINTS
Large (1,000 + 1 lbs)	20
Moderate (50-999 lbs)	10
Low (<50 lbs)	5

Process incidents exceeding 75 points result in the following actions:

● Key lessons are documented, analyzed, and shared company-wide.
● Corrective actions are taken at other locations to prevent a recurrence.
● Key lessons and identified PSM elements provide input to corporate PSM technology leadership teams.

Incidents scoring more than 130 points are classified as serious process incidents. Incidents in this category are subjected to in-depth root cause analysis, and the follow-up from company-wide sharing is stringent.

5.5.6   Results
What are the results? Table 5.13 (see page 112) presents March 1996 lost workday case (LWC) statistics. Audit scores for 1994 and 1995 are shown in Table 5.14 (see page 112), as well as identified PSM elements in 1995 process incidents. Here are some highlights of results:

● No major property damage/losses (greater then $1 million) since 1992.

● 1995 was the first year in recent history in which there were no process-related LWCs in high hazard processes. Previously (1971-1994), the company had averaged 7 process-related LWCs per year.

Table 5.13   *Safety performance - DuPont C&S plants (March 1996)*

PLANT	COUNTRY	EXPOSURE HOURS SINCE LAST LWC†	CALENDER DAYS SINCE LAST LWC	NUMBER OF EMPLOYEES
1	USA	38,558,460	8,466	716
2	USA	36,301,852	2,339	2,447
3	USA	27,053,586	8,876	89
4	USA	20,492,732	7,636	545
5	Taiwan	20,378,544	6,100	84
6	Mexico	19,469,382	6,555	508
7	USA	19,149,604	12,736	57
8	USA	19,029,160	2,007	1577
9	Belgium	18,858,804	6,337	721
10	UK	18,446,958	3,190	951

† LWC = lost workday case

Table 5.14   *Audit scores and identified PSM elements*

	PROCESS SAFETY MANAGEMENT (PSM) SECOND-PARTY AUDIT SCORES (%)		PSM ELEMENTS IDENTIFIED IN PROCESS INCIDENTS OF >75 POINTS (%)
	1994	1995	1995
Total	82	84	
Management	76	84	
Process technology	85	82	9
Process hazards analysis	80	81	7
Management of change - technology	82	79	4
Operating procedures	80	85	18
Training and performance	85	91	19
Contractor safety	90	90	2
Management of change - personnel	69	73	0
Incident investigation	85	87	3
Emergency response	88	90	2
PSM auditing	69	74	2
Quality assurance	85	90	6
Mechanical integrity	74	85	21
Management of subtle change	78	84	3
Pre-startup review	81	84	3

The results in part reflect actions taken as a result of the leading indicator - process safety metrics. Therefore, the DuPont conclusion is: safety is good business!

5.5.7   References

1.   Scott I A P, 1993, *How to measure success in process safety*, Presented at an IPSG technical meeting, San Francisco, USA.

2.   Scott I A P, 1994, DuPont's approach to managing process safety, in Cacciabue P C, Gerbaulet I and Mitchison N (eds), *Safety Management Systems in the Process Industry*, Proceedings of CEC Seminar, Ravello, Italy, October 1993, 98-104 (European Commission, Luxembourg).

3.   OSHA, 1992, *Process Safety Management of Highly Hazardous Chemicals*, Title 29, Code of Federal Regulations, Part 1910.119 (Occupational Safety and Health Administration, Department of Labor, Washington, DC, USA).

4.   *Common Position (EC) No 16/96 on Council Directive 96/.../EU on the control of major accident hazards involving dangerous substances*, 1996 (Council of the European Union, Brussels, Belgium).

5.   *ISO 9000: Quality Management and Quality Assurance Standards — Guidelines for Selection and Use*, 1987 (International Organization for Standardization).

6 Measuring the outputs: practical examples

The previous three chapters were concerned with examples from three areas of safety management inputs: plant and equipment, systems and procedures, and people. Measuring these inputs is required in order to introduce a way of giving assurance that the absence or reduction of harm or loss is due to a systematic safety and risk management approach which includes the plan-do-check-improve cycle of advanced management systems. Injuries, illnesses, losses and so on have to be measured as well, however, since through these measurements it will eventually be clear whether the inputs were provided successfully.

Therefore, this fourth chapter with examples is focused on the 'bottom line' of safety performance: on measuring the outputs. Section 6.1 presents output measurements at Norsk Hydro. It discusses a number of output indicators, such as the lost-time injury rate and the total recordable injury rate, and contains various ways of presenting the results of output measurements graphically. It also briefly describes three different auditing methods at Norsk Hydro, and concludes by giving an indication of the benefits from the company's safety activities. Next, a contribution from Dow in section 6.2 is more specifically concerned with the issue of the costs and benefits of safety. The question is addressed whether there can be an economic return from a safety programme. The emphasis is on the quantification of savings. Data for Dow are presented together with data for the general and chemical industry, for various indicators such as fatalities, days away from work cases and long-term disability claims. The conclusion is that it is possible to find the required information for assessing the economic benefit from a successful safety management system.

6.1   Output measurements
W Bjerke, Norsk Hydro

Norsk Hydro started its safety performance measurements by focusing on negative output (failures). This is a reactive approach which indicates what is going wrong. The company is now heading increasingly in a more proactive direction to find indicators which reveal what can be done in advance, in order to prevent things going wrong:

● Which weaknesses occur in the safety management system? What is not functioning?
● Which preventive actions can be taken to control the risks?
● How are actions followed up?
● How to use output measures to evaluate results and trends, in order to determine whether the company is doing things right, is doing the right things, and achieves its goals and targets?

There is also a trend towards trying to find positive indicators (for example, how many positive observations relative to the total number of observations). So far, the company has few indicators on process safety.

The performance indicators which are presently used at site, divisional and corporate levels include:

● Fatal accident rate (FAR, for own employees and contractors).
● Lost-time injury (LTI) and LTI-rate (for own employees and contractors).
● Lost-time injury severity (LTIS) rate (rate of lost workdays due to injuries).
● Total recordable injury rate (LTI + restricted work + other medical treatment).
● Major incidents, divided into fires, explosions, accidental releases, work accidents and other (for example, near-misses with high potential).
● Place of incident, divided into process, storage, pipe, transport and other.
● Operation before incident: process, maintenance and other.
● Expenses for all insured losses: material damage, production loss, etc.
● Percentage of sick leave due to accidents and illnesses.
● Environmental emission figures, deviations from permits, etc.

The indicators are used regularly by line management and safety, health and environment personnel to draw attention to all types of failures (human error, unsafe acts, unsafe conditions and generally all deviations from standards and good practice). This is done to motivate for higher quality, as part of a continual improvement process.

Some of the indicators are also connected to concrete goals and targets — for example, those set annually at each organizational level. Goals and targets should be linked to concrete action plans (programmes) which indicate how to achieve the goal, and the plans should be monitored.

Other elements of structured performance measurement include regular reporting to the different organizational levels, communication of incident causes (both direct causes and root causes) and analysis of developments.

The following definitions are used in this section:

● Incident: an unintended work-related event (or sequence of events) that may result in an undesirable consequence.
● Accident: an incident which results in injury to persons and/or damage to property.
● Near-miss: an incident which, in other circumstances, could have become an accident.

6.1.1 Monitoring activity

A 'Form for recording and reporting of accidents and near-misses' is the input to any incident monitoring activity (see Figure 6.1, pages 116 and 117). As part of this, incidents are classified in one of the following three categories: minor, medium and major (see Table 6.1, page 118). The particular category indicates which investigation activity and resources will be used to find direct and root causes. For near-misses, the potential of the incident determines the classification.

	Classification: INTERNAL
HYDRO	K13-1-CTS-02 Appendix 1

Revision: 2	Sign.: *Ole Rømer*	Date: 1995-01-01	Page 1 of 2

Title: FORM FOR RECORDING AND REPORTING OF
          ACCIDENTS AND NEAR-MISSES

Division:	☐ Accident
Unit:	☐ Near-miss
Unit code:	Date of incident: ....

INCIDENT DESCRIPTION

Classification:	Type:	Place:	Exposure of persons:
☐ Major	☐ Explosion/Fire	☐ Process	☐ Extreme temp./radiation
☐ Medium	☐ Accidental emission	☐ Storage	☐ Hazardous liquid/gas
☐ Minor	☐ Work accident	☐ Piping	☐ Electricity
	☐ Other (describe)	☐ Transport	☐ Fall
	....	☐ Other (describe)	☐ Hit/struck by
		....	☐ Crushing
Emergency actions:		**Operation:**	☐ Sharp/pointed object/tool
☐ External authority alert		☐ Production	☐ Splinters/dust
☐ Internal alert		☐ Maintenance	☐ Overstraining
		☐ Construction	☐ Other (describe)
		☐ Other (describe)	....

Dominant material type involved : .... Quantities: .... kg/liters
            released as ☐ Liquid   ☐ Gas   to ☐ Air   ☐ Water   ☐ Ground
Component/equipment involved: ....
Ignition source: ....

Description of incident/sequence of events, including any other materials involved:

CAUSES

Classification:
☐ Technical failure of equipment   ☐ Inadequate job instruction
☐ Inadequate protection of equipment   ☐ Wrong application of job instruction
☐ Physical working environment   ☐ Insufficient training/competency
☐ Other:......................................   ☐ Inadequate use of personal protective equipment
                                               ☐ Unknown cause

Description of causes (mention direct and indirect causes):

Ref. Investigation report:

Figure 6.1   *Form for recording and reporting of accidents and near-misses*

```
┌─────────────────────────────────────────────────────────────────┐
│                                        ┌──────────────────────┐   │
│  HYDRO                                  │ Classification: INTERNAL │
│                                         K13-1-CTS-02  Appendix 1   │
│  Revision: 2       Sign.: Ole Bruaas    Date: 1995-01-01   Page 2 of 2 │
```

Title: FORM FOR RECORDING AND REPORTING OF
          ACCIDENTS AND NEAR-MISSES

Unit ....   Date of incident:...........................

OCCUPATIONAL INJURY/LOSS (CONSEQUENCES)

Number of injuries/fatalities: Dominant type of injury: (If more than one
                                      person is injured give the number by the
                                      relevant type):

Employees: Hired personnel:
_ Fatality   _ Fatality   _ Head/neck/mouth   _ Hip/knee/leg
_ Lost time injury*) _ Lost time injury _ Eye   _ Ankle/foot
_ Restricted work   Other contractors: _ Shoulder/arm   _ Respiratory system
_ Medical treatment _ Fatality   _ Hand/finger   _ Other internal injuries
_ First aid   _ Lost time injury _ Chest/stomach   _ Serious multiple injury
                                        _ Back   _ Other: ....

*) Number of lost working days due to injury: ....
Description of injuries (for near-misses describe potential):

MATERIAL DAMAGE/PRODUCTION LOSS (CONSEQUENCES)

Material damage: KNOK.....................   Production loss:
                                              No. of hours: ....
                                              Capacity reduction: .... % by value

Description of damage (for near-misses describe potential damage/loss):

CORRECTIVE/PREVENTIVE MEASURES PROPOSED, incl. changes in management

Completed by: ....   Date: ....
Updated by: ....   Date: ....

To be distributed according to instructions from the division management.

Figure 6.1 continued

6.1.2   Typical safety indicators
Fatal accident rate
FAR is calculated as follows:

FAR = Number of fatalities x 100 x 10^6 / hours worked

Fatal accidents are rare events, and a five-year moving average is often used to illustrate trends. The FAR for employees and the FAR for contractors are recorded separately or combined.

Table 6.1   *Classification of accidents and near-misses*

MAJOR	MEDIUM	MINOR
Fatality or lost-time injury leading to hospitalisation	Lost-time injury not requiring hospitalisation, injury requiring medical treatment and injury leading to restricted work	First-aid injury
Material damage greater than NOK† 2 million	Material damage from NOK 500,000 to NOK 2 million	Material damage less than NOK 500,000
Accidental emission with potential for long term or substantial short-term environmental damage and/or that is reported to state authorities	Accidental emission with potential for limited short-term environmental damage and/or that is reported to local authorities	Accidental emission in excess of permit with insignificant potential for environmental damage

+ NOK = Norwegian Kroner

Note: The values in the table are designed to serve as guidelines. In case of doubt the higher classification should be chosen.

Lost-time injury rate
LTI rate is calculated as follows:

LTI rate = Number of lost-time injuries x 10^6 / hours worked

The LTI rate is probably the most common safety performance indicator in industry. Norsk Hydro's definition of lost-time is absence of one day or more, not including the day of injury. Others use injuries per 200,000 hours instead of 10^6 hours and/or three days or more of absence. Thus, it is important to define how the LTI rate is measured to be able to benchmark.

Lost-tme injury severity rate
LTIS rate is calculated as follows:

LTIS rate = Number of days lost due to injuries x 10^6 / hours worked

The LTIS rate indicates the severity of the lost-time injuries. LTIS rate/LTI rate gives the number of days lost per injury and indicates the 'average' severity of each injury.

Total recordable injury rate
TRI rate is calculated as follows:

TRI rate = (Number of LTI + RWC + MTC) x 10^6 / hours worked

RWC is the number of restricted work cases, and MTC is the number of other medical treatment cases. The TRI rate measures the rate of all injuries which involve medical treatment. First aid cases are not included in the TRI rate, but are of course recorded as incidents. (Although the number of first aid cases is recorded, frequency rates are normally not calculated for first aid cases.)

Examples of medical treatment cases include injuries which require treatment by a doctor or a by nurse under the guidance of a doctor, such as surgery, stitching of wounds or the use of prescription medicines. Injuries which require first aid treatment, such as simple treatment of wounds or rinsing of eyes, are not regarded as medical treatment, even when carried out or administered by a doctor.

Number and costs of process-related incidents
The number and calculated loss for fires, explosions, accidental releases and so on indicates the level of control of risk (or process safety level). These can be calculated for each type of material damage, and also for combinations of different types of material damage. The business interruption (production loss) is often also calculated separately. Material damage includes direct cost, replacement value and any direct third-party loss and clean-up costs.

Transport/distribution accident rate
Incidents to people, environment and property during transportation by ship, road and rail should be recorded. A rate can be calculated — for example, per million tonnes of goods transported — for each transport medium, and also for combinations of different transport media. The distribution accident rate (DAR) is calculated as follows:

DAR = Number of transport accidents / tonnes transported by ship/road/rail

6.1.3   Occupational safety and health indicators
The main indicator which is used for occupational safety and health is percentage of sick leave due to illness and injuries. It gives the percentage of absence due to illness or injury. Some companies also calculate the absenteeism rate, the rate of new cases of sick leave and injuries and the number of occupational diseases.

Absenteeism rate
The absenteeism (ABS) rate is calculated as follows:

ABS rate = Number of work days lost due to illness or injury x 10^6 / hours worked

Rate of new cases of sick leave and injuries
The rate of new cases of sick leave and injuries (NC) is calculated as follows:

NC rate = Number of new cases of illness or lost-time injury x 10^6 / hours worked

The number of cases of occupational diseases is calculated and reported regularly on a local level.

6.1.4   Presentation of results
A monthly report is issued throughout the company, on paper and by E-mail, in which group results, division results and the results of each site are presented for the LTI rate (monthly result, year-to-date, last twelve months, last year, goal this year, days since last injury). Table 6.2 gives an example. The same is done for percentage of sick leave and NC rate. Major accidents are described in terms of what happened and the associated consequences. Figure 6.2 (see page 122) shows a way to present LTI rate and LTIS rate as a percentage reduction for consecutive years compared with a reference year. Figure 6.3 (see page 122) gives trends in reduction of LTI rate for employees and contractors, whereas Figure 6.4 (see page 123) presents the TRI rate for employees. Figure 6.5 (see page 123) shows the development of serious accidents.

6.1.5   Analysis of measurements
The measurements and indicators are used to analyze trends, and often tell management on which items to focus. Weaknesses are commented upon and strengths praised. Results are observed and compared with concrete goals, and management are held accountable for safety performance. The comparisons between sites act as a sort of 'management by shame' activity and have a remarkable influence on the process of continuous improvement. Nobody wants to be left behind.

6.1.6   Follow-up on performance measurements; future development
At sites where results are bad it is common practice to conduct safety audits, in order to check the implementation of the safety management system and identify strengths and weaknesses. Three different systems are described below. The first is an example of a simple method, shown in Figure 6.6 (see page 123), for general health, environment and safety (HES) issues, in which management was asked questions by an audit team about the implementation and functioning of 19 general issues/elements in their safety management system. Scores are given from 0 to 5 (from bad to excellent). This audit system has been used at division and site level for general safety elements.

        For a more detailed and specific process safety management (PSM) audit, the following 11 elements, related to process safety, have been distinguished and used in a similar way:

1.   Process safety information.
2.   Process safety studies.

Table 6.2 *Hydro safety record: site statistics (monthly report)*

April 1996

Division/Group	Site			Number of employees	Lost-time injuries this month	Lost-time injuries per million hours worked					Days since last LTI
						This month	Year to date	Last 12 months	1995	Goal 1996	
INDUSTRIAL CHEMICALS											
Hydro Chemicals	Hydro Chemicals Ltd		GB	18	0	0	0	0	0	0	912
	Hydro Kemi	A	SE	33	0	0	0	0		0	
	Hydrowax		NO	6	0	0	0	0	0		
	Ind. Chemicals Oslo	A	NO	20	0	0	0	0	0		
	Pardies		FR	194	1	39	19	7	0	0	5
	Chemtech		DE	122	1	59	13	9	4	0	12
	Hydro Chemicals Norge		NO	37	0	0	0	16	30	0	254
	SUM			430	2	34	12	7	4	0	
Hydrogas	Hydrogas AS	A	NO	27	0	0	0	0	0		
	Hydrogas France		FR	11	0	0	0	0	0		
	Hydrogas Holland BV		NL	16	0	0	0	0	0		
	Hydrogas Ltd		GB	28	0	0	0	0	0		
	Hydrogas Polska	A		5	0	0	0	0	0		
	Ceylon Oxygen		CE	234	0	0	0	2	2		331
	Hydrogas Norway		NO	203	0	0	0	3	5		309
	Hydrogas Denmark		DK	141	0	0	0	7	22		140
	KWD GmbH		DE	103	0	0	0	10	10		167
	SUM			768	0	0	0	4	7		
Others	Hydelko		NO	41	0	0	0	0	14		455
	SUM			41	0	0	0	0	14		
	SUM INDUSTRIAL CHEMICALS			1239	2	11	4	5	6		
MAGNESIUM											
Mg Metal	Offices Mg	A		23	0	0	0	0	0	0	
	Bécancour		CA	360	1	19	13	7	5	3	19
	Porsgrunn-Mg		NO	495	0	0	4	8	7	3	42
	SUM			878	1	8	8	7	6	4	
Other Mg	Magnesiumgesellschaft		DE	48	0	0	0	14	13	0	302
	SUM			49	0	0	0	12	10	4	
	SUM MAGNESIUM			927	1	7	7	8	7	4	

A = administration (fully or partly)

3.   Operating procedures.
4.   Safe work practices.
5.   Modifications (process/equipment).
6.   Quality control and maintenance.
7.   Competence and training.
8.   Investigation and reporting of accidents.
9.   Emergency planning and response.
10.   Pre-start-up safety reviews.
11.   Inspections and auditing.

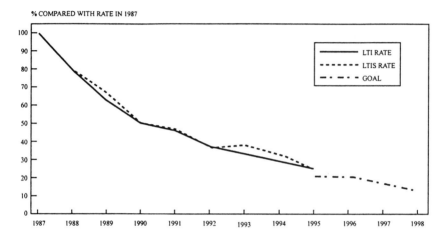

Figure 6.2   *Reduction in LTI rate and LTIS rate*

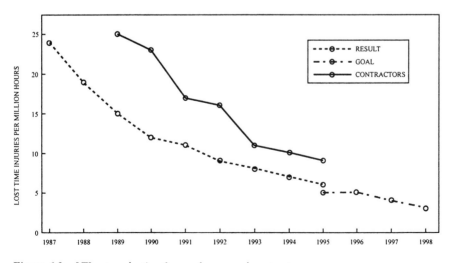

Figure 6.3   *LTI rate reduction for employees and contractors*

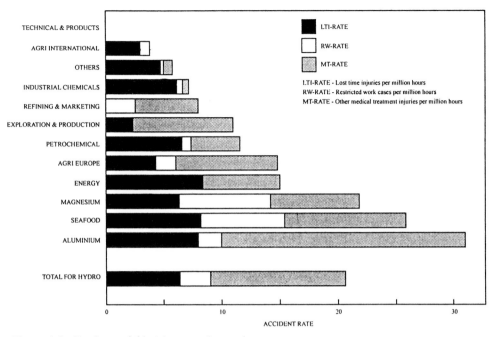

Figure 6.4 *Total recordable injury rate for employees*

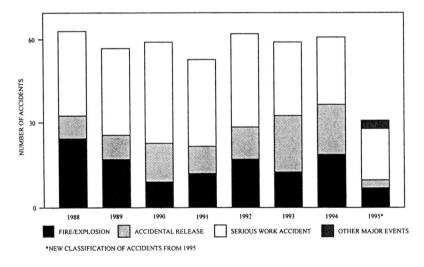

Figure 6.5 *Development of serious accidents*

These two types of audits are more qualitative than quantitative, and currently a more quantitative approach is under development, together with Veritas. The system is developed from the International Safety Rating System (ISRS), with the intention to integrate the assessment of occupational health and safety, process safety and environmental affairs.

This audit, in checking the adequacy of the safety management system, also includes an evaluation and score of performance results and physical conditions. It concludes with an overall assessment (see Table 6.3).

The system audit results are presented for 12 system elements which are shown in Table 6.4. Each element is based on detailed sub-elements with a score for each question.

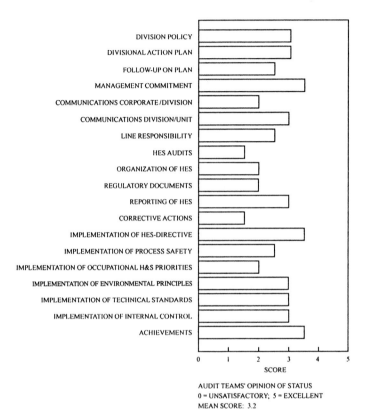

AUDIT TEAMS' OPINION OF STATUS
0 = UNSATISFACTORY; 5 = EXCELLENT
MEAN SCORE: 3.2

Figure 6.6   *Health, environment and safety (HES) profile corporate audits: typical profile 1992*

Table 6.3   *Hydro safety rating system: overall score*

SITE: _____   DATE: _____

ELEMENT	WEIGHT	SCORE									
System audit	50%										
Physical condition evaluation	25%										
Performance results	25%										
		0	10%	20%	30%	40%	50%	60%	70%	80%	90%
Overall score											
Division		THIRD DIVISION			SECOND DIVISION			FIRST DIVISION			

0   40%   75%   100%

Table 6.4 *Hydro safety rating system: score from system audit*

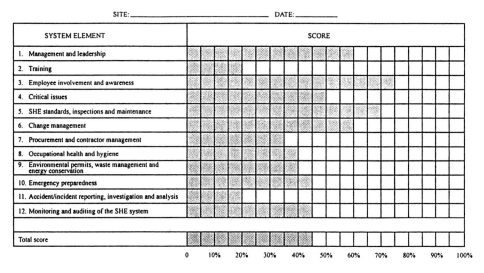

6.1.7 Conclusion

The overall conclusion is that work on safety pays. This can clearly be seen in the survey of expense savings in Hydro from 1985 to the present day, as given in Table 6.5.

Table 6.5 *Measuring the benefit*

1985	1994
Fatal accident rate (FAR)* = 3.9 in the period 84/88 (average of 2 fatal accidents per year for Hydro employees)	Fatal accident rate (FAR) = 0.35 in the period 90/94 (average of 1 fatal accident every 5 years for Hydro employees)
1500 - 1600 Lost-time injuries (average 5 - 17 days)	365 Lost-time injuries. Direct cost reduction alone amounts to NOK† 50 million per year.
NOK 175 million in direct material costs as a result of fires and explosions, lost production not included	NOK 50 million in direct material costs as a result of fires and explosions

* FAR = fatal accidents per 100 million hours

† NOK = Norwegian Kroner

6.2 Can there be an economic return from a safety programme?
R Gowland, Dow Chemical

In the past the concept of safety has been adopted as a credo without attempting to see how much it might cost. In other words, there were few attempts to quantify the cost a given programme might cause. There were also few attempts to see if there are any savings to be made. Industry has been working at making improvements without a clear idea of cost and benefit.

Indeed, an understanding of the link between the two has often been lacking. This is not a criticism; it is a statement of fact in most cases. As industry has moved to better and better performance, looking at Figure 6.7 could illustrate that it has perhaps moved from the horizontal part towards the right hand end where the returns for investment are becoming less.

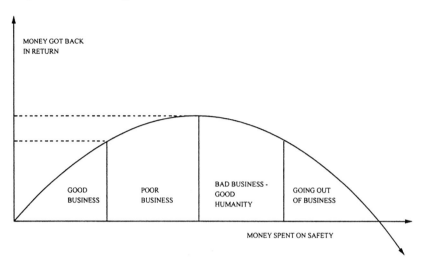

Figure 6.7   *Returns for investments on safety — taken from Reference 1*

These thoughts are to some extent reflected in the following statements which may appear in discussions on safety:

● Safety costs money.
● Industry spends the money but still has accidents.
● Industry reacts to incidents by adding safety features and restrictions.
● The human factors are not well understood.
● If safety was a 'stand alone' business, no one would invest in it.

Thus, it is becoming much more important to measure the costs and benefits of safety. Perhaps it is easier to measure the costs than it is to measure the benefits. In this section it is intended to pay more attention to the latter, although even the first is not easy. Success in making the cost which has been associated with safety and environment accepted as a normal cost of doing business would be real progress.

6.2.1   The costs of safety

The costs of safety are built up in the traditional cost centres and accounts. This data should be readily available. However, if industry has succeeded in making safety a part of everyone's job, some of the costs may not be identifiable. Dow has found this difficult.

It appears that some costs are apparent and some are hidden. The costs for the employees of a safety department are apparent. The costs for some other issues are partly hidden. Examples of hidden costs are:

1.   Extra safety built into design, operation and maintenance for manufacturing plant.
2.   The apparent extra time taken by every employee for safety activities.
3.   Protective clothing and equipment.

Given the existence of hidden costs, it is necessary to have a closer look at these in order to be able to get an idea of the costs of safety.

Examples of the first category of hidden costs — extra safety built into design, operation and maintenance for manufacturing plant — are:

- larger safety factors on structural steel;
- sprinkler and deluge systems;
- redundant instrumentation;
- cost of inspection and maintenance.

Many of these have been built into design standards and have 'disappeared'. It seems as if industry is doing it but is not quite sure why. In some cases industry practice is the result of 'reflex reaction' to a major incident. This needs to be challenged. An alternative to trying to find the cost is to make sure that only the cost which is really necessary to manage the risk is incorporated.

In many cases it is possible to use a risk and consequence basis for design. A simple 'what if' or Hazop study can give consequences of system failure. These can be compared with simple matrices drawn up to give a design requirement. Unless this approach is taken, there is a danger of overspending. An example (from Dow Chemical) for the selection of the correct type of instrument redundancy to meet increasing level of process risk is given in subsection 6.2.5. Using a risk-based approach like this, the cost of safety disappears and the cost is a cost of doing business responsibly and to criteria set and agreed within the company. In many cases, the degree of redundancy required for critical instrument systems has been reduced whilst still giving the necessary overall reliability. This is particularly the case where analogue instruments are used. These can be monitored for their condition throughout their operating life by carefully written software.

Application of the risk-based approach in inspection and maintenance is discussed in section 3.4 on condition measurement.

The second category of hidden costs — the apparent extra time taken by every employee for safety activities — is difficult to quantify. It is probably not necessary, however, since the ingenuity of people will be able to make the safe way of doing a job the most efficient. Work practices which introduce more restrictions and inefficiencies are rarely sustainable with the work force and usually result in violations and accidents. Behavioural safety can only work if the working methods are optimized.

As to the third category of hidden costs — protective clothing and equipment — experience at Dow shows that the careful selection and use of protective clothing is not an added cost unless personal discipline is a problem. Bad habits and abuse of equipment are the biggest problems. These may impose cost, but have nothing to do with safety.

For Dow, the information shows that the largest costs are in categories 1 and 2.

6.2.2   Quantification of savings from a successful safety management system

The UK Health and Safety Executive has attempted to lead the way with the document 'The costs of accidents at work'[2]. Dow has looked at the same subject to see whether the effort put into safety management and safety management systems has resulted in lower incident-related costs. Can this be quantified? Dow's answer is yes, and the company is now projecting savings which will accrue from successfully meeting the targets set for the year 2005.

Dow has attempted to quantify as follows by benchmarking with others and checking its historic performance. (All data are USA and Dow USA, unless stated otherwise. Corresponding statistics for Dow Europe are slightly better than for Dow USA.)

Fatalities

The average annual fatality rate per 50,000 workers is 4.35 for the general industry, whereas it is 0.00 for Dow. This is an important category of potential savings: the cost of a fatal accident has been estimated at US$6.1 million.

Days away from work (DAW) cases

The average DAW rate, for 1990-1993 and per 200,000 hours, was 2.16 for the general industry, 0.58 for the chemical industry and 0.45 for Dow.

The DAW cost per 100 workers was US$13,000 for the general industry, US$1300 for the chemical industry and US$1200 for Dow.

Long-term disability

The percentage of employees with long-term disability claims, for 1991-1994, was 2 for the general industry and 1 for Dow. The number of occupation-related new claims per 100 employees per year was 0.3 for the general industry and 0.1 for Dow. Each long-term disability claim can cost up to US$3.9 million.

Workers' compensation payments

Workers' compensation payments are payments into a compulsory federal insurance fund for compensation to injured workers. The 1994 cost per US$100 payroll was US$3.3 for the general private industry, US$7.22 for the chemical manufacturing industry in Texas and US$0.3 for Dow.

Fatal motor vehicle accidents, whilst on company business

The 1985-1994 national average of fatal motor vehicle accidents whilst on company business was 32, whereas it was 2 for Dow.

OSHA penalties

US Occupational Safety and Health Administration (OSHA) penalties are legal citations for violations of OSHA rules. The 1990-1994 chemical industry average was US$1,381,000. The amount for the 'worst in group' was US$4,294,800, whereas it was US$20,350 for Dow.

A US study (by Fry and Lee) found that the stock value of a given company was affected by penalties. The perception was that additional funds would be needed to respond to the inadequacy revealed.

Losses due to fires
The 1985-1994 USA petroleum industry average for losses due to fires was US$0.1 per US$100 of insurable value, whereas it was less than US$0.02 per US$100 of insurable value for Dow.

Insurance premiums
As a result from performance improvement, Dow has reduced its insurance premium by 25% for property loss and business interruption over the last few years. It now costs US$27 million per year (Dow global). Based on historical premiums, the cost without the performance improvement is estimated at US$36 million.

Other
There are some categories which are more difficult to quantify. Examples include image, and upset for neighbours, the public, authorities, lawyers and the media. However, the effect on major corporations after major incidents can be determined.

6.2.3   Conclusion
This has been an attempt to give some ideas of the economic benefit which a successful safety management system can bring. It shows that the required information can be found. Each company will have different standards and objectives, and therefore the savings will not be the same for everyone. Nevertheless, the process and headings are common to all.

The economic benefit of a successful safety management system is only an argument to support what is, in fact, a much greater moral imperative. All the same, it is very useful to be able to explain to commercial colleagues what the economic realities are.

6.2.4   References
1.   Kletz T A, 1992, *Hazop and Hazan — Identifying and Assessing Process Industry Hazards*, 3rd edition (Institution of Chemical Engineers, Rugby, UK).

2.   Health & Safety Executive, 1991, *The Costs of Accidents at Work*, Health and Safety Series Booklet HS(G)96 (HMSO, London, UK).

6.2.5   Appendix: defining the critical instrument system
For the design and testing of critical instrument systems, a 'route map' (Figure 6.8) may be used. As part of the procedure, the matrix as shown in Table 6.6 (see page 131) is suggested as a guide. To use the matrix, evaluate the potential consequences of the instrument system failure and classify it into one of four possible hazard classes.

The approach may be used for new systems as well as for existing systems. For new systems, this allows designers to specify suitable instrument systems (via IEC 1508: see Table 6.7, page 132) to meet the suggested 'integrity level' for each class of hazard and the preferred inspection frequency. For existing systems, it allows appropriate inspection intervals to

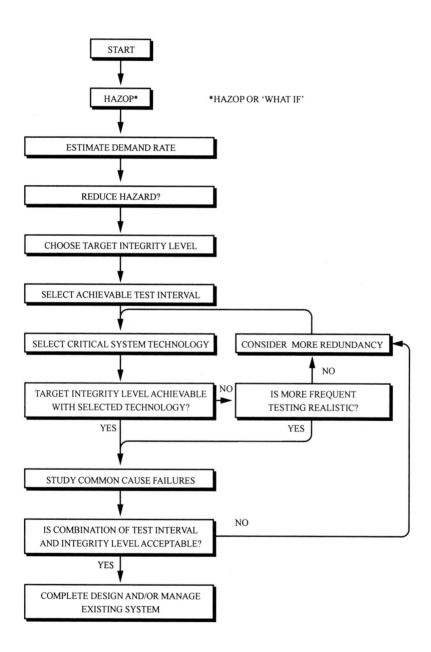

Figure 6.8   'Route map' for defining the critical instrument system

be set by using the guidance on reliability, redundancy and inspection intervals which is provided for Dow in its Loss Prevention Principle 15.4 on Critical Instrument Systems or in Table 6.7 (see page 132).

An obvious point comes out of the theory used for the inspection intervals — the maximum set must be realistic, and where very large intervals are indicated it shows that this level of redundancy is not needed.

The methodology is that the company defines the risk level in terms of injury, cost and so on which it wants to avoid, and then sets its criteria accordingly. In the example presented here, which is taken from Dow Loss Prevention Principle 15.4 on Critical Instrument Systems, the reliability and degree of redundancy for an instrument system controlling a hazard can be set.

Table 6.6 *Risk matrix*

HAZARD CLASS (POTENTIAL CONSEQUENCE)	'INTEGRITY LEVEL'
● Multiple fatalities on or off site	
● Permanent environmental damage to an area > 5 hectares of a natural habitat off site	4
● Property damage or business interruption > $ 800 million	
● A fatality on or off site, injury (resulting in hospitalisation) to a member of public off site	
● Environmental incident resulting in clean up off site or a fine	
● Any event leading to the need to evacuate members of the general public	3
● Property damage or business interruption > $ 50 million and < $ 800 million	
● Fire or explosion leading to property damage off site	
● DAWC (day away from work case)	
● Any incident resulting in local public being told to shelter indoors	
● Environmental incident which involves off site clean up or which could contaminate ground water	2
● Property damage or business interruption > $ 1 million and < $ 50 million	
● Fire which requires off site fire fighters to control	
● Minor injury as a result of chemical exposure or RMTC (recordable medical treatment case)	
● Environmental incident where contamination is confined to site and where recovery is complete in 1 year, or is reportable to a government agency	1
● Property damage or business interruption > $ 50,000 and < $ 1 million	
● Fire which is controlled by site personnel	

Table 6.7 *Requirements for final elements according to IEC 1508 Part 2, Table A5 (oo = out of)*

Safety integrity level (SIL)	Final element configuration	Diagnostic cover per channel	Off-line proof test interval (TI)	Mean time to false trip (MTTF spurious) on-line/off-line repair
3	*Single*	*High*	*≤ 5 months*	*4.4 years*
	Dual (2 oo 2)	*High*	*≤ 2 months*	*30188 years*
	Dual (1 oo 2)	*None*	*≤ 4 months*	*4.4 years*
	Dual (1 oo 2)	*Low*	*≤ 10 months*	*4.4 years*
	Dual (1 oo 2)	*Medium*	*24+ months*	*4.4 years*
	Dual (1 oo 2)	High	24+ months	*4.4 years*
	Triple (2 oo 3)	*None*	*≤ 2 months*	*> 160 years*
	Triple (2 oo 3)	*Low*	*≤ 6 months*	*> 136 years*
	Triple (2 oo 3)	*Medium*	*24+ months*	*> 147 years*
	Triple (2 oo 3)	High	24+ months	> 1367 years
2	*Single*	*Low*	*≤ 1 month*	*5.5. years*
	Single	Medium	≤ 5 months	4.6 years
	Single	High	24+ months	4.4 years
	Dual (2 oo 2)	Medium	≤ 2 months	> 4458 years
	Dual (2 oo 2)	High	24+ months	> 4097 years
	Dual (1 oo 2)	None	≤ 13 months	4.4 years
	Dual (1 oo 2)	Low	24+ months	4.4 years
	Dual (1 oo 2)	Medium	24+ months	4.4 years
	Dual (1 oo 2)	High	24+ months	4.4 years
	Triple (2 oo 3)	None	≤ 7 months	> 50 years
	Triple (2 oo 3)	Low	≤ 19 months	> 49 years
	Triple (2 oo 3)	Medium	24+ months	> 147 years
	Triple (2 oo 3)	High	24+ months	> 1367 years
1	Single	None	≤ 5 months	8.8 years
	Single	Low	≤ 13 months	5.5 years
	Single	Medium	24+ months	4.6 years
	Single	High	24+ months	4.4 years
	Dual (2 oo 2)	None	≤ 2 months	> 476 years
	Dual (2 oo 2)	Low	≤ 6 months	> 404 years
	Dual (2 oo 2)	Medium	24+ months	> 436 years
	Dual (2 oo 2)	High	24+ months	> 4097 years

Note 1:   For safety integrity level 4 a detailed quantitative hardware analysis is highly recommended, therefore no SIL 4 architecture is presented. SIL 4 is not acceptable at Dow facilities

Note 2:   Off-line proof test intervals of more than 24 months are not explicitly calculated. For all tables a mean time to repair of 8 hours is assumed

Note 3:   The italics area is only allowed for low fault count components

7 *Conclusion*

The management of safety is an issue which is receiving more and more attention in the chemical process industries. Many companies have developed and are implementing safety management systems. This trend is reflected in recent regulatory developments. In 1996, the so-called Seveso Directive of 1982[1], which was strongly focused on the technical side of safety, is replaced by the Seveso II Directive[2], which puts more emphasis on safety management. The new directive contains requirements for a major accident prevention policy and specific safety management system aspects. Similar developments in the USA led to similar requirements in the US Occupational Safety and Health Administration (OSHA) Regulation on process safety management[3].

Within safety management, the attention is expanding from implementation to include measurement: from safety management systems to safety performance. Currently, there is not a large body of literature on safety management systems and safety performance, and much of the information is scattered in conference proceedings[4-6].

Following a Commission of the European Communities (CEC) seminar on safety management systems in the process industry[7], the European Process Safety Centre published a book on safety management systems[8]. That publication went beyond the level of generic safety management system concepts and was mainly concerned with presenting typical examples from industry.

The purpose of this book is to broaden the issue of implementing safety management systems by adding a perspective on measuring their performance. The approach taken is similar to that of the previous publication. Therefore, this book starts by presenting a generic framework for performance measurement, but its emphasis is on current — and developing — practice in industry: it is particularly focused on actual examples.

The framework for performance measurement presented in this book consists of two dimensions: areas of safety management inputs and monitoring activities. Three areas of safety management inputs are distinguished: plant and equipment, systems and procedures, and people. Within each of these three areas, various kinds of monitoring activities can be distinguished, ranging from regular, and often frequent, inspections through periodic, 'in-depth' assessments to overall management audits. In three subsequent chapters, examples of the various monitoring activities are presented for each of the three areas of safety management inputs. A separate chapter is devoted to the outputs: the bottom-line of safety performance.

It must be emphasized that the approach taken in this book has limitations. Each of the examples is only a part of the overall system of the company in question, but the examples together do not constitute an overall system of monitoring and measurement since they derive from different overall systems. Moreover, designing and implementing safety performance measurement in a particular company has to be tailored to the needs and characteristics of the company in question.

These needs and characteristics are concerned with at least two aspects. Firstly, the emphasis in managing safety should constantly move between the different areas of safety management inputs in order to realize continuous improvement. Since different companies will be in different stages as far as their emphasis is concerned, their needs will also differ. Secondly, there exist clear differences between companies in terms of organizational culture, and this perspective is an important one to address in achieving effective safety management[9].

Thus, although the final word has not yet been said about safety performance measurement, it is hoped that this book will advance the application of safety performance measurement techniques in industry and thereby will contribute to an increased understanding of safety management as well as to higher levels of safety in industry.

7.1   References

1.   *European Council Directive on the major accident hazards of certain industrial activities*, 1982, Directive 82/501/EEC (European Community, Brussels, Belgium).

2.   *Common Position (EC) No 16/96 on Council Directive 96/.../EU on the control of major accident hazards involving dangerous substances*, 1996 (Council of the European Union, Brussels, Belgium).

3.   OSHA, 1992, *Process Safety Management of Highly Hazardous Chemicals*, Title 29, Code of Federal Regulations, Part 1910.119 (Occupational Safety and Health Administration, Department of Labor, Washington DC, USA).

4.   *International Process Safety Management Conference and Workshop*, San Francisco, USA, September 1993 (Center for Chemical Process Safety, American Institute of Chemical Engineers, New York, USA).

5.   Wicks P J, Cole S T and Berman J (eds), 1994, *Operational Safety*, Proceedings of Joint CEC-ESReDA Seminar, Lyon, France, October 1993 (European Commission, Luxembourg).

6.   Mewis J J, Pasman H J and de Rademaeker E E (eds), 1995, *Loss Prevention and Safety Promotion in the Process Industries*, Proceedings of the 8th International Symposium, Antwerp, Belgium, June 1995 (Elsevier Science).

7.   Cacciabue P C, Gerbaulet I and Mitchison N (eds), 1994, *Safety Management Systems in the Process Industry*, Proceedings of CEC Seminar, Ravello, Italy, October 1993, (European Commission, Luxembourg).

8.   EPSC, 1994, *Safety Management Systems: Sharing Experiences in Process Safety* (Institution of Chemical Engineers, Rugby, UK).

9. van Steen J F J and Brascamp M H, *The importance of organizational culture in achieving effective safety management* (Department of Industrial Safety TNO, Apeldoorn, The Netherlands) (to be published).